ELIMINATING THE UNIVERSE

Logical Properties of Natural Language

ELIMINATING THE UNIVERSE

Logical Properties of Natural Language

EDWARD L KEENAN

NEW JERSEY · LONDON · SINGAPORE · BEIJING · SHANGHAI · HONG KONG · TAIPEI · CHENNAI · TOKYO

Published by

World Scientific Publishing Co. Pte. Ltd.

5 Toh Tuck Link, Singapore 596224

USA office: 27 Warren Street, Suite 401-402, Hackensack, NJ 07601

UK office: 57 Shelton Street, Covent Garden, London WC2H 9HE

British Library Cataloguing-in-Publication Data

A catalogue record for this book is available from the British Library.

ELIMINATING THE UNIVERSE
Logical Properties of Natural Language

ISBN 978-981-4719-83-4

For any available supplementary material, please visit
https://www.worldscientific.com/worldscibooks/10.1142/9779#t=suppl

Desk Editor: Benny Lim

Printed in Singapore

This work is dedicated to:
Aryeh Faltz and Jonathan Stavi
for sharing their beautiful minds with me early on,
and to John Corcoran,
who taught me that logic was a mode of linguistic analysis.

Eliminating the Universe:
Logical Properties of Natural Language

This work streamlines, generalizes, and hopefully makes more accessible the natural language semantic work I have pursued for the last 35 years. I'm taking the opportunity to put much of that work in a unified setting, and to focus on results rather than proofs which can be consulted in the original sources, if desired. This work was originally intended to be a collection of reprints of my articles, but I found that uninspiring. Still, I focus here on my own work and make no attempt to summarize the current state of the field.

In my endeavors, I have benefited enormously from collaboration with Leonard M. Faltz, Jonathan Stavi, Lawrence S. Moss, Dag Westerståhl, and Edward P. Stabler. Especially in the early stages, I relied on their mathematical competence which was well beyond my own. To this list I should add Johan van Benthem, Marcus Kracht, and Uwe Mönnich, for while we have not coauthored any works, my interactions with them and their work prompted and critiqued my own very significantly. I would also like to thank the linguistics department at UCLA for having tolerated my wanderings in the boolean forest; and I owe a debt as well to the mathematics and the philosophy departments here for permitting me to present my work at their colloquia on many occasions. Deep thanks also to my dissertation advisor John Corcoran who helped me to see that mathematical logic was the linguistic analysis of mathematical languages (Elementary arithmetic, Euclidian geometry, ...).

Lastly, three thanks of a different sort: first to my lovely wife, Dr. Carol Archie, who has had to put up with untold hours of my uffish thought with nary a medical application in sight. Second, to George Lakoff, who generously gave me a boost when I had just finished my dissertation — namely, he actually read the thing, a three valued logic replete with theorems which purported to model presupposition in natural language, not at all the sort of work that generative grammarians were prepared to read at that time. And third, to Noam Chomsky whose early works, especially (1956, 1959), led me to think I could discover interesting things about the mind without doing any work, just sit on my duff, ponder my language and prove some theorems. Of course things didn't turn out as simple as that, but this work appears to have brought me full circle.

Reader Alert This work focuses on the boolean structure of grammatical categories of English expressions. It is a simple fact, indeed a rather mundane, even boring, fact that most "content" expressions — predicates, arguments, modifiers, sentences, permit the productive formation and interpretation of boolean compounds in *and, or, not,* and *neither... nor....* It turns out, as the diligent reader will see, that the semantic structure of categories (Chapter 8) runs much deeper than merely permitting the interpretation of boolean connectives (*and, or, ...*). Yet it is this mundane, superficial level, that, ultimately, is the most arresting.

As competent speakers of English, we accept as well formed and interpret unproblematically such diverse boolean compounds as: *an intelligent **but not** very industrious student*; *She **neither** praised **nor** criticized each student*; *We'll **either** come early **or** leave late*; *Some undergraduate **but not** Bill will represent us at the meeting*; ***Not** every senior and **not** every freshman skipped class*; *He lives **neither** in **nor** near New York City*; *Most **but not** all freshmen think they're clever*; *She works **either** carefully **or** rapidly **but not** both*; ***Neither** are whales mammals **nor** are spiders insects*; *I doubt whether there is life on Mars **but not** whether there is water there.* Equally boolean compounds may have multiple occurrences within the same sentence: *Bob **and either** Sue **or** Ann will arrive early, take the tickets **and** clean up afterwards; **Neither** AL **nor** Ann criticized **both** Ed **and** Ted.*

I take the syntactic ubiquity of the boolean connectives as evidence that their meaning is not specific to any given category. Sentences, predicate phrases, quantified arguments, and modifiers are meaningful in different ways, but boolean compounding applies in all cases. So the meaning of the boolean connectives reflects more how we look at things than how things are, properties of mind we might say, or "laws of thought", in the felicitous phrase of George Boole (Chapter 1). And they are perhaps not quite unique. *Only* and *also* are modestly ubiquitous and they also to my thinking represent how we look at something relative to other things — like contrastive stress (Association with Focus; Jackendoff 1972).

For this reason then we have a short Chapter 1 which cites some of Boole's master work, published as *The Laws of Thought* (1854). Within the tradition of modern linguistics we have documented our history reasonably well. But

Boole is certainly not among our cited precursors. And indeed he was a mathematician not a linguist, but he had a remarkably modern conception of the tight fit between language and mind, and in effect anticipates several points in the foundation of modern linguistics.

Reading this Book   The brief Chapter 1 on George Boole is a stand-alone chapter. Occasional reference is made to it later but the application of boolean notions to linguistic analysis in later chapters is self-contained. Chapter 2 provides a brief overview of mathematical logic as a mode of linguistic analysis and is relied on later. Often in later chapters I have repeated some notation and definitions so that the chapters are approximately self-contained. As indicated above, each chapter draws on my earlier work, but in no case have I actually taken over, much less quoted the earlier works. The intent is to streamline the presentation and I have cited my sources accordingly. Only on a few occasions do I actually give an explicit proof, and that mostly later in the presentation where I think it might be either quick and easy or not so readily available. Loosely, the complexity of the material treated increases as we move through the chapters. So I am hopeful that linguists with only a limited exposure to mathematical logic will find material of interest in the early chapters. By Chapter 12, Eliminating the Universe, the chapter that justifies the title of the book, I show how to replace the primitives of a standard model with one lacking a universe of objects in favor of one in which properties, common noun denotations, are taken as a primitive. This, I claim, is useful in the representation of properly intensional evaluative adjectives (*skillful*). Whether it has any other utility or philosophical interest is open.

Contents

Overview of this Work   I address myself here primarily to linguists and logicians, but also to computer scientists involved with natural language processing (NLP).

To linguists, I hope to enrich our field with the use of mathematical techniques adapted from logic, enabling us to discover generalizations that were beyond our ken until we learned how to express them. Galileo said it best: *Mathematics is the language of science.* To logicians, I hope that drawing attention to novel semantic phenomena in natural language will provide more raw material and motivation for extending the substantive results you have achieved on mathematical languages. And to those in NLP I hope that the syntactic and semantic regularities I note among expressions in diverse categories may be helpful in areas like machine learning, data mining and artificial intelligence in which understanding entailment and paraphrase relations among natural language expressions is needed.

Chapter 1

The Linguistic Prescience of George Boole

We begin with a brief discussion of the insightful and prescient vision of the relation between language and mind that George Boole (1815–1864) presents in his 1854 magnum opus:

AN INVESTIGATION

of

THE LAWS OF THOUGHT

ON WHICH ARE FOUNDED

THE MATHEMATICAL THEORIES OF LOGIC AND PROBABILITIES

The present work draws on boolean structure extensively, though Boole's *Laws*, as I will call the above cited work, is not a normal historical citation in formal semantics or generative grammar. In the former domain we expect to see foundational work such as Frege (1893), Tarski (1931), and Montague (1970). In the latter, Chomsky (1986, Ch 1) notes sources ranging from Plato and Aristotle through Descartes, the Port Royal grammarians, von Humboldt, John Stuart Mill, and more recently the linguist Otto Jesperson (1924).

The name *George Boole* is notably missing from these source lists, and not without reason. Boole was primarily a mathematician. His *Laws* presents the first instantiation of what we know today as Boolean Algebra, a well developed subfield in mathematics with extensive application in logic and information sciences. Boole was not an incipient linguist. There is no danger of confusing his *Laws* with the Port Royal Grammar (Arnauld and Lancelot 1975; see Chomsky 1966/2009). But Boole shares with contemporary, and earlier, linguistic theory the view that there is a tight relation between language and mind. The opening sentence in the *Laws* begins:

> The design of the following treatise is to investigate the fundamental
> laws of those operations of the mind by which reasoning is performed;
> to give expression to them in the symbolical language of a Calculus, . . .
> (p. 1).

Boole feels "That Language is an instrument of human reason, . . . " (p. 26);
"The elements of which all language consists are signs. . . " (p. 27) and ". . . in
investigating the laws of signs, . . . , the immediate subject of examination is
Language, with the rules which govern its use;" (p. 27). And Boole continues
here with evidence for the tight interconnection between language and mind:

> Nor could we easily conceive, that the unnumbered tongues and dialects
> of the earth should have preserved through a long succession of ages so
> much that is common and universal, were we not assured of the existence
> of some deep foundation of their agreement in the laws of the mind itself
> (p. 27).

Is this not the modern linguists' Universal Grammar (UG)! Compare Chomsky (1975, p. 4) who agrees that "language is a mirror of the mind" understood as: *More intriguing, to me at least, is the possibility that by studying
language we may discover abstract principles that govern its structure and
use, principles that are universal by biological necessity and not mere historical accident, that derive from mental characteristics of the species.* So the
grammars of different languages (Japanese, Malagasy, English, . . .) are all
special cases of UG, which is biologically determined.

Given Boole's investigation of language and mind, it is perhaps not surprising that he makes other observations compatible with current linguistic
thinking. Here are a few examples, with just a quick comment regarding
their relevance in modern syntax and semantics:

> "*Definition* — A sign is an arbitrary mark, having a fixed interpretation,
> and susceptible of combination with other signs in subjection to fixed
> laws dependent upon their mutual interpretation" (p. 28). Does this not
> anticipate Frege's compositional semantic interpretation?

> "The Romans expressed by . . . "civitas" what we designate by . . . "state."
> But both they and we might equally well have employed any other word"
> (p. 28). The arbitrariness of the sign is, again, usually attributed to de
> Saussure (1916).

Boole distinguishes between primary and secondary propositions — the latter being ones that relate propositions to each other. For example (p. 58)
"The sun shines" is primary. . . ", "If the sun shines the earth is warmed" is
secondary. He then points out that some sentences built with what today
we call binary boolean connectives (*and, or, if-then*) are primary, not being
"resolvable into" secondary ones. An example (p. 59) of linguistic interest is:

"*Men are, if wise, then temperate* cannot be resolved into *If all men are wise then all men are temperate*".

Now, after the explosion of linguistic work created by Chomsky's *Syntactic Structures* in 1957, many linguists were concerned to formulate operations that syntactically derived one expression from another (or both from a given abstract source). It seemed at first that many pairs of sentences derived from the same source had the same logical meaning: *Ted smokes and drinks* was derived from *Ted smokes and Ted drinks*, and they are indeed true in the same states of affairs (models). Another operation derived *Ted wants to leave* and *Ted wants Ted to leave* from the same source. But it wasn't long before linguists noticed that these derivational operations failed to yield logical paraphrases with quantified phrases: *Some of the men expected to be drafted* is not true in the same models as *Some of the men expected some of the men to be drafted* (Jackendoff 1972, p. 205).

Once again Boole is ahead of the pack.

Lastly, lest we overstate Boole's linguistic prescience, we note that generative grammar has several concerns that were simply not shared by Boole. One was the productivity of the derivational processes used to generate expressions. They were recursive, so from a finite set of words (a lexicon) we derive infinitely many expressions: *He knows a doctor, He knows a doctor who knows a doctor, He knows a doctor who knows a doctor who knows a doctor, ...* In contrast Boole did illustrate recursion using *and* and *or*: *Wealth consists of things transferrable, limited in supply, and either productive of pleasure or preventive of pain* (p. 65) in which a disjunction is embedded within a conjunction. But iterativity per se is not discussed and we find no "unbounded" examples, such as *John and either Sam or both Sue and either ...* or no remarks about deriving infinitely many expressions with finite means.

Chapter 2

Logic as Universal Grammar

In this chapter, I provide the motivation for taking standard logics — Sentential Logic, First Order Logic, Higher Order Logic — as models of linguistic analysis. Our presentation is informal, designed to be intelligible and hopefully interesting to scholars who are not familiar with the mathematical perspectives that logic brings to linguistic analysis. The later chapters presuppose some basic mathematical proficiency, manipulating functions and relations.

Originally I thought primarily of linguists as the audience for this chapter. My exposition of first order logic is hardly novel. But perhaps the linguistic perspective I take is somewhat novel. A recent book *Three Views of Logic: Mathematics, Philosophy, and Computer Science* 2014, by Loveland et al., presents logic from three substantive perspectives, none of which is the linguistic one taken here.

1. Sentential Logic (also called Propositional Logic; 'Logic' may be replaced by 'Calculus' in these expressions).

First and higher order logic define classes of languages, as we see shortly, but Sentential Logic is basically just a single language (with some notational variation of interest). We begin with it as it is syntactically and semantically the simplest of the logics we discuss, and it illustrates in a simple way several basic logical concepts.

In general a logic specifies three things: (1) syntactically, a set of expressions (a *language*), (2) semantically a definition of *truth in a model* in terms of which *entailment* is defined, and (3) a definition of *proof*, a syntactic characterization of the entailment relation. We consider proofs only in this chapter.

SENTENTIAL LOGIC

The Language (SL)   Syntactically, SL is a set of expressions called *formulas* (or *sentences*) constructed in two steps: first, an infinite set $\{P_1, P_2, \ldots\}$ of syntactically unanalyzable *basic* formulas is stipulated. Second, SL itself is defined to be the minimal set of expressions containing the basic formulas and closed under appropriate combinations with *and*, *or*, and *not*. Specifically, the rules of the grammar of SL are:

(1)   a.   All basic formulas P_i are in SL (noted $P_i \in$ SL).

   b.   If p and q are in SL (noted: $p, q \in$ SL) then $(p\,\mathrm{and}\,q) \in$ SL. p and q are the *conjuncts* of $(p\,\mathrm{and}\,q)$, itself called a *conjunction* of sentences. For '$(p\,\mathrm{and}\,q)$' we often write '$(p \,\&\, q)$'.

   c.   If $p, q \in$ SL then $(p\,\mathrm{or}\,q) \in$ SL. p and q are its *disjuncts* and $(p$ or $q)$ itself called a *disjunction*.

   d.   If $p \in$ SL then $(\mathrm{not}\,p) \in$ SL.   ($\mathrm{not}\,p$ is called a *negation*; The parentheses are not needed).
   (For the record these rules are given more formally in an Appendix to this chapter).

   For example $(P_3 \,\&\, \mathrm{not}(P_2\,\mathrm{or}\,\mathrm{not}\,P_3))$ is a formula.   The tree diagram below summarizes the argument that this is so (F = 'Formula').

(2)

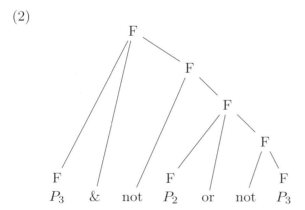

Some common alternate notations: $\wedge$ is often used for *and*, $\vee$ for *or*, and $\neg$ for *not*.   We reserve these symbols for other purely semantic purposes later. Basic formulas are often called *atomic* but we use 'atomic' in a purely semantic sense later. We have used infix notation, writing the binary boolean connectives *and* and *or* between the formulas they combine with. In prefix (Polish) notation these connectives are written formula initially and often abbreviated A, O, and N. In prefix notation the formula in (3) would be:

(3) $AP_3NOP_2NP_3$

In either infix or prefix notation we observe (when the definitions are given more rigorously) the following facts:

Fact 1. Each $p \in$ SL is either a basic formula or derived in just one way by combining with *and*, *or*, and *not*.   □

So each formula is either a basic formula, a conjunction, a disjunction, or a negation of formulas, but never more than one of these simultaneously.

Fact 2. Each $p \in$ SL contains just finitely many basic formulas and p itself is a finite sequence of symbols. (The set SL is infinite, so it is an infinite collection of finite objects.)   □

The Semantics of SL   The semantic interpretation of SL formulas depends on just one primitive notion — truth. And even here our primary concern is relative truth: if p is true does it follow that q is? So we want to define for all formulas p, q in SL what it means to say that p *entails* q (which means the same as p *logically implies* q, or q *logically follows* from p).

We emphasize that entailment is a *semantic* relation. If we can show that two sentences do not entail exactly the same sentences then they clearly differ in meaning. Here is a simple case:

(4)   a.   Ted can both read and speak Korean

      b.   Ted can either read or speak Korean

      c.   Ted can read Korean

Clearly (4a) entails (4c). If (4a) is true in some state of affairs then (4c) must also be true in that state. But imagine a state in which Ted is an illiterate Korean speaker. Then (4b) is true but (4c) false, so (4b) does not entail (4c). So (4a) and (4b) clearly differ in meaning (and since syntactically they differ by *and* vs *or* we suspect that the meaning difference between them is due to that difference).

Note that it is easy to imagine a state of affairs in which (4b) and (4c) are both true. Our entailment claim is simply that from (4b) alone we are not justified in inferring (4c) since in some states of affairs (4b) is true and

(4c) is not. But inferring (4c) from (4a) is justified since (4c) is true in *all* states of affairs in which (4a) is.

So our semantic analysis of SL must characterize the notion "state of affairs", or as we will now say, *model*. For SL this notion is quite simple (but is more complex in logics with a richer syntax).

Naively first, an English sentence such as *Both Ted and Ned came to the lecture but Ned didn't stay long* is built from simpler ones by *and* and *not* (*n't*), and whether it is true in a state of affairs is determined by truth in that state of the simpler sentences we have formed the conjunction and negation of. The truth of the simple sentences depends on the state of affairs, but not on other sentences. In different states of affairs they may have different truth values.

Our semantics for SL has nothing of substance to say about the truth of basic sentences $P_1, P_2, \ldots$ Different assignments of truth values to them represent different ways the world is. So a *model* for SL is any function m from the set BF = $\{P_1, P_2, \ldots\}$ of basic formulas, into the two element set of truth values {False, True} which we represent as $\{0, 1\}$, using, as is common, 1 for True and 0 for False. So if m and m' are different models they assign different truth values to at least one basic formula — for some P_i, $m(P_i) \neq m'(P_i)$.

The truth value of non-basic formulas depends on (is a function of) those of the basic formulas it is built upon: (P_1 & not P_2) should be interpreted as 1 (true) in any model m in which $m(P_1) = 1$ and $m(P_2) = 0$. And if m' is a model with $m'(P_1) = 0$ or $m'(P_2) = 1$ then (P_1 & not P_2) should be interpreted as 0 (false).

To capture this dependency, we extend a model m to an *interpretation* m^* of the whole of SL as follows:

Definition 1. For each model m, m^* is that function from SL into $\{0, 1\}$ which maps each basic sentence P_i to the truth value m maps P_i to; it maps a conjunction of formulas to 1 iff it maps each conjunct to 1; it maps a disjunction to 1 iff it maps at least one of the disjuncts to 1; and it maps (not p) to 1 iff it maps p to 0.

And we may now define entailment ($\models$) and the related notion *valid*:

Definition 2.

> a. For all $p, q \in$ SL, p *entails* q, noted $p \models q$, if and only if for all models m, if $m^*(p) = 1$ then $m^*(q) = 1$.

b. For all $p, q \in \text{SL}$, p is *logically equivalent* to q, noted $p \equiv q$, iff $p \models q$ and $q \models p$.

c. For all $q \in \text{SL}$, q is *valid*, noted $\models q$, iff for all models m, $m^*(q) = 1$. A synonym for *valid* is *logically true*.

d. For S a subset of SL and $q \in \text{SL}$, $S \models q$ iff for all models m, $m^*(q) = 1$ if for all $p \in S$, $m^*(p) = 1$.

One shows easily that logically equivalent formulas have the same truth value in each model — either both true or both false.

Fact 3 (Coincidence Lemma). For all $p \in \text{SL}$, if models m and m' assign the same truth values to the basic formulas that occur in p then, $m^*(p) = m'^*(p)$.

Fact 3 says that the truth of a formula only depends on the truth of the basic formulas that occur in it.

Fact 4. In SL validity is decidable.

That is, there is an algorithm which, when applied to any $p \in \text{SL}$, says Yes if p is valid, and No if p is not valid.

The procedure mentioned in Fact 4 is the method of *truth tables*: For any p there is a number n (Fact 2) such that there are just n basic formulas that occur in p (some perhaps many times). So there are 2^n ways of assigning them truth values. Write down the 2^n sequences of truth values under the basic formulas and then in each case compute the truth value of the successive subformulas using Definition 1 until we have a value for the entire formula. If it is 1 for all the assignments it is valid, otherwise it is not. For example:

$$(5) \quad (\text{not } P_2 \text{ or } (\text{not } P_3 \text{ or } P_2))$$

1	1	1
1	0	1
0	1	0
0	0	0

Check the second line: not P_2 is false since P_2 is true. And not P_3 is true, as is P_2, so (not P_3 or P_2) is true, so the entire disjunction in (5) is true. Similarly the other lines always compute to 1, so (5) is valid.

Also, given our semantics, we can now extend the syntax of SL by adding new connectives without extending expressive power since, as shown below, we can replace the formulas they build with logically equivalent ones we already have. Below we add *if-then* ($\rightarrow$), the biconditional ($\leftrightarrow$), and *nei-ther... nor...* (the Sheffer stroke, noted $(p|q)$).

(6) a. We write $(p \rightarrow q)$ to abbreviate (not p or q).

 b. We write $(p \leftrightarrow q)$ to abbreviate $((p \rightarrow q)\ \&\ (q \rightarrow p))$.

 c. We write $(p|q)$ for (not p & not q).

Note that we could have defined *or* in terms of & and *not*: *(p or q)* ≡ *not(not p & not q)*. In fact we could have defined <u>all</u> the connectives in terms of *neither... nor...*: *not p* ≡ *neither p nor p*, $\overline{(p\ or\ q)}$ ≡ *not neither p nor q*, and *(p and q)* ≡ *not(not p or not q)*. One may wonder just how many truth functional binary connectives one *could* define. The answer is 16. (Why?)

Proofs in SL   A proof here is basically a finite sequence of sentences of SL designed to show convincingly that the last sentence in the sequence is entailed by (logically follows from) the earlier sentences in the sequence which are marked in some way as premises. To define a proof we define syntactically a finite number of *rules of inference* and show they are *sound* (*correct*). That is, if a rule sanctions deriving some q from some premises $p_1, \ldots, p_k$, then q is true in all models in which all the premises are true.

For example, one rule is Conjunction Elimination: from $(p\ \&\ q)$ derive p (and also q), and we write $(p\ \&\ q) \vdash p$. That is, if $(p\ \&\ q)$ is a line in a proof from some premises we can add p and we have a proof of p from the same premises. This rule is obviously sound, as, trivially, $(p\&q) \models p$, all p, q in SL. A more interesting rule is Modus Ponens: From $\{p, (p \rightarrow q)\} \vdash q$. That is, if both p and $(p \rightarrow q)$ are lines in a proof then we can add q to it and the result is a proof of q (from the same premises). By truth tables we compute that for all models m^*, if $m^*(p) = 1$ and $m^*(p \rightarrow q) = 1$ then indeed $m^*(q) = 1$, so Modus Ponens is sound.

Linguistically, proofs are a kind of text and rules of inference are ways of building texts from texts. Further, the sort of natural deduction systems we have partially adumbrated here are actually more than a mere sequence of formulas, they have some internal structure specific to proofs. Here are, informally, some examples.

Suppose we want to prove a conditional claim $(p \rightarrow q)$. We might begin by overtly assuming p, marking it as an assumption, and then with sound rules of inference derive q, and then infer $(p \rightarrow q)$ from no assumptions, that is, blocking off the original assumption of p. A related example is proof by *reductio ad absurdum*: to prove some p, we assume not p, derive a q we know to be false (say a contradiction (r & not r)), and then infer p from no assumptions.

A slightly trickier case is what is called *disjunctive syllogism*. We want to show, for a certain choice of p, q, r, that (p or q) entails r. So we first assume p and derive r, then we assume q and derive r, and then we close off these two subproofs and infer r just from (p or q).

We see then a study of proofs would be an interesting study in text linguistics, but it is not a study we pursue in this work. We only note the following important theorems about SL:

Notation. For S a subset of SL, noted $S \subseteq$ SL, and $q \in SL$, $S \vdash q$ means there is a proof of q from premises in S.

And since our ways of building proofs, the rules of inference, are sound, we infer:

Theorem 1 (*Soundness*). For all $S \subseteq$ SL, and all $q \in$ SL, if $S \vdash q$ then $S \models q$.

The other logics we look at also all have sound deduction systems (rules of inference). But Second (and Higher) Order Logic fails the converse of Soundness, called Completeness:

Theorem 2 (*Completeness*). For all $S \subseteq$ SL, and all $q \in$ SL, if $S \models q$, $S \vdash q$.

So Completeness of a logic says that whenever a sentence follows from some premises, then there is a proof of that sentence from those premises. Since proofs are syntactically defined, they constitute a syntactic characterization of the (semantic) relation entailment.

Completeness has a corollary of some linguistic interest: namely the truth of no sentence can depend on infinitely many pieces of independent information. For SL, this may seem obvious, since the truth of any q in SL just depends on that of the finitely many basic formulas that occur in it. But this claim also holds for First Order Logic where no such decomposition is guaranteed to hold. Here is the theorem in its standard mathematical form:

Theorem 3 (*Compactness*). For all $S \subseteq$ SL, and all $q \in$ SL if $S \models q$ then there is a finite subset S' of S such that $S' \models q$.

Theorem 4 (*Interpolation*). For p, q neither true in all models nor false in all models, if $p \models q$ then there is a formula r, called an *interpolant*, such that

$p \models r$ and $r \models q$ and the atomic formulas occurring in r are just those that occur in both p and q.

So Interpolation is a relevancy theorem. That some p non-trivially entails some q just depends on what they have in common.

Reflections on Sentential Logic   SL is obviously quite restricted in expressive power. Basically it just gives us the meaning of the truth functional connectives (*and*, *or*, *not* and others definable in terms of them) applied to sentences. Surely this will not take us very far in terms of natural language semantics. And indeed we need much more than SL has to offer.

But SL does illustrate in a simple way what a compositional semantics looks like, and it shows what semantic definitions, like entailment look like, and it shows how this semantic relation can be syntactically characterized.

SL also has two specific properties that make it unsuitable as a model for natural language semantics. The first, which is not shared by the richer logics we consider, is that it contains infinitely many sentences with no proper parts. If English were like this how would we ever learn what they meant? It is not hard to see that English has infinitely many sentences (as well as infinitely many expressions of other categories): *Ed knows a doctor, Ed knows a doctor who knows a doctor, Ed knows a doctor who knows a doctor who knows a doctor...* But we can figure out what each of these sentences means if we know what their lexical items mean and how they are put together. But the basic sentences of SL have no proper parts whose meaning can be learned, so no way to infer their meaning in general.

And second, the sentences of SL are syntactically unambiguous — each is either basic or built in a unique way with *and*, *or*, and *not*. But natural languages are structurally ambiguous, even with regard to the scope of *and* and *or*. *Ted and Ned or Jed* might mean the same as *Either both Ted and Ned or Jed* or *Both Ted and either Ned or Jed*. So the coordinate initial *either* and *both*, serve to disambiguate the scope of the coordination, just as coordinate initial conjunctions do in Polish notation. Can we always disambiguate syntactically ambiguous sentences in English?

Interim conclusion   SL is clearly limited in expressive power compared to English, though surprisingly as we will see, the boolean structure it builds on encompasses much more than SL. But first let us briefly present First Order Logic, extended variants of which are widely used in representing natural language semantics.

First order logic (FOL)

FOL specifies a class of languages, unsurprisingly called *first order* languages. These have been, and are, widely used in representing semantic properties of natural language. We focus on these properties here in informally presenting FOL. For clear, basic but fully formal introductions to FOL, see Ebbinghaus et al. (1984) and Enderton (1972). A formal, more linguistically oriented treatment is Keenan and Moss (2016).

A first order language L forms boolean compounds of formulas as in SL. But it no longer has infinitely many syntactically unanalyzable basic formulas. It forms *atomic* formulas from predicate symbols and their argument expressions, and *quantified* formulas combined with *variable binding*.

To consider what is new in FOL we consider some examples of formulas from the first order language of elementary arithmetic and compare them with some English sentences that have properties expressible in FOL. We provide some semantic motivation for first order syntax as we go along.

Consider first an atomic sentence such as $(1 > 0)$ in arithmetic. It says that the number one is greater than the number zero. The symbols '1' and '0' are called *individual constants* and are treated semantically as denoting objects in the *universe* of a model for the language of arithmetic. So part of what a model of a first order L must do is specify a universe E of *individuals* that we use L to talk about. Individual constants of L will be semantically interpreted as elements of this universe. Different interpretations of L may have different universes. The language of elementary arithmetic is a little unusual in that it has a "standard model", one whose universe is $\mathbb{N}$, the set of natural numbers $0, 1, 2, \ldots$. (But it also has non-standard models whose universes properly include $\mathbb{N}$.)

In the atomic sentence $(1 > 0)$ the symbol '>' is a P_2, a two place predicate symbol. It combines with two *terms*, such as individual constants, to form a formula.

We can analyze an English sentence such as *Ted respects Ned* in a way similar the one we use for $(1 > 0)$. We think of the expressions *Ted* and *Ned* as denoting objects in the universe of things we use English to talk about, and we may think of *respects* as similar to '>'. Namely it semantically relates two individuals, denoted by *Ted* and *Ned* in this example, just as the relation denoted by '>' relates the numbers denoted by '1' and '0'. In general an n-place predicate, a P_n, requires n occurrences of terms to form an atomic formula, and n is called the *arity* (*degree*, *valence*) of the P_n. P_n's denote n-ary relations, ones which hold of n individuals. We note though that P_ns in FOL are not marked for tense (present, past, etc.) and so do not distinguish

Ted respects Ned from *Ted respected Ned.* In our later logical analysis of English we shall also not treat tense but in relevant examples we will hold tense constant so that differences in semantic interpretation of interest to us do not depend on tense variation.

Syntactically an atomic formula built from a P_n and n terms is usually written with the P_n initial, followed by the terms. The exception is when $n = 2$, as above, when we often write the P_2 between the terms. As we use the term first order language we understand that the equality symbol $=$ is part of the language. It functions syntactically as a P_2 and is a logical constant, always interpreted as absolute identity. Thus in a first order language a formula of the form $a = b$ is always understood to mean that a and b are interpreted as the same member of the universe.

Terms consist of more than individual constants. Complex terms are built by combining n-place function symbols with n terms. For example $(1+0)$ is a term in arithmetic, where '$+$' is an F_2, a 2-place *function symbol*, and of course '1' and '0' are terms. So in general an F_n combined with n terms is a term. (An individual constant can be thought of as a 0-place function symbol — it combines with zero terms to make a term, as it already is one.)

Lastly, FOL provides denumerably many *individual variables* $x_1, x_2, \ldots$ as terms. So $(x_2 > 0)$ is an atomic formula in arithmetic. In referring to expressions in a first order language, the boolean connectives *and*, *not*, etc., the quantifier symbols, and the equality predicate are called *logical constants* — their semantic interpretation does not vary from model to model, whereas the other predicate and function symbols that distinguish one language from another are called *non-logical constants*.

A *model* for a first order L is a pair (E, m) where E is a (usually non-empty) set, the *universe* of the model, and m is a "lexical" denotation function mapping each individual constant to an element of E, each n-ary function symbol to a function from E^n into E, and each n-ary predicate symbol to a function from E^n into $\{0, 1\}$. (E^n is the set of n-ary sequences $(b_1, b_2, \ldots, b_n)$ of elements of E.) Linguists often write $[\![\cdot]\!]$ rather than m.

To define an interpretation of L in a model (E, m) we will need some way to say how individual variables take their denotations. The functions that map them to elements of E are independent of the function m, since, as we see below in interpreting quantified formulas, we must be able to let the denotations of variables vary holding the denotations of constants, function and predicate symbols unchanged. So we define an *assignment* (of values to the variables) to be a function from the variables $x_1, x_2, \ldots$ into E.

Now let us define the three types of formulas in a first order L and say informally how they are semantically interpreted. *Atomic* formulas consist of an n-place predicate symbol P (typically) followed by n terms: $Pt_1, \ldots, t_n$. Next *boolean compounds* of formulas are those formed with *and, or*, and *not* and other connectives definable in terms of them, as in SL. And finally quantified formulas, those of the form $Qx\varphi$, where Q is a quantifier symbol, either $\forall$ or $\exists$, x is a variable, and φ a formula. $\forall$, an upside down A, is the *universal quantifier* and formulas of the form $\forall x\varphi$, read as *for all* x φ, are said to be *universally quantified*. $\exists$, a backwards E, is the *existential quantifier*. $\exists x\varphi$ is read as: *For some* x φ or *There is an* x *such that* φ. Such formulas are *existentially quantified*.

Now, an *interpretation of L relative to a model (E, m) and all assignments g* is a function which maps each expression in L to an appropriate denotation, as follows: (1) Each "lexical item" d (individual constant, n-place function or predicate symbol) denotes $m(d)$, whatever m says it denotes; (2) each variable x denotes $g(x)$, whatever g says it denotes, g any assignment; (3) each complex term $Ft_1, \ldots, t_n$ denotes whatever object $m(F)$ applied to the n-tuples of denotations of $t_1, \ldots, t_n$ is; (4) each atomic formula $Pt_1, \ldots, t_n$ denotes whatever truth value $m(P)$ and g map the n-tuple of denotations of $t_1, \ldots, t_n$ to; (5) boolean compounds: $(\varphi \,\&\, \psi)$ denotes 1 relative to any g iff φ denotes 1 relative to g, and ψ denotes 1 relative to g (and analogously for disjunctions and negations, per SL). (6) Finally, $\forall x\varphi$ denotes 1 relative to g iff φ itself denotes 1 relative to all assignments g' that differ at most from g in what they assign to the variable x. And $\exists x\varphi$ denotes 1 relative to g iff for some g' differing from g at most at x, φ denotes 1 relative to g'. And we now define entailment relative to an assignment g, $\models_g$, by:

Definition 3. For all formulas φ, ψ, all assignments g, $\varphi \models_g \psi$ iff for all models (E, m) all interpretations relative to (E, m) and g, if φ is interpreted as 1 then ψ is interpreted as 1.

We may note that the interpretation of a formula like $(x_2 > 0)$ depends crucially on the choice of assignment: if $g(x_2) = 7$ the formula is true, but if $g(x_2) = 0$ it is false. However if φ is a *closed* formula (called a *sentence*), meaning that each occurrence of variable x in φ is within a constituent of the form $Qx\psi$ then for each model, φ is interpreted as the same truth value at all assignments. So entailment between sentences depends solely on the model: $\varphi \models \psi$ iff for all models (E, m), if φ is interpreted as 1 (at any assignment) then so is ψ. Similarly a set S of sentences entails a sentence ψ iff ψ is interpreted as 1 in all models in which all the sentences in S are interpreted as 1.

Major Properties of FOL   The notion of *proof* is defined for arbitrary first order Ls and is provably *sound*: if there is a proof of a sentence φ from a set S of sentences then S models φ. *Completeness*, also holds: if $S \models \varphi$ then there is a proof from (finitely many) premises in S to φ. So compactness also holds, and an appropriately restated version of interpolation holds — the interpolant now just contains non-logical constants (such as P_ns) common to the non-trivial sentences in the entailment relation.

As in SL, the completeness of FOL provides a syntactic characterization of the (semantic) entailment relation since $S \models \varphi$ iff $S \vdash \varphi$. But due to some remarkable theorems by Per Lindström 1969, FOL is basically a maximal logic in which this is true. Just about any way of increasing the expressive power of FOL loses most of its nice properties (like completeness).

FOL differs from SL with regard to *decidability*. There is no mechanical procedure which will tell us for an arbitrary first order L whether a sentence is valid (true in all models) if it is and whether it isn't if it's not. Since proofs are finite sequences syntactically defined, we can enumerate the infinitely many proofs over the formulas in a first order L, and to test whether some φ is valid we can start cranking out proofs. But after any finite amount of time, if we haven't yet given a proof of φ (from no premises), we don't know if it is because there isn't one or we just haven't gotten to it yet. So validity in FOL is said to be *semi-decidable* or *recursively enumerable*.

Some merits of representing English semantics in FOL:

1. We have already mentioned that FOL allows us to discriminate predicates and arguments. This is a starting point for a richer analysis of English. Thus linguists distinguish semantic subclasses of P_ns. Contrast *Ted respects Ned* with *Ted criticizes Ned (often)* — in the latter, Ted performs an activity of which he is the Agent and may be held responsible. But *respects* determines a state, not an activity. Similarly, *criticizes* naturally allows modification with frequency adverbs, whereas as *Ted respects Ned often* is unclear in meaning. And coordinate with states and mere activities linguists would distinguish *accomplishments* like *draw a circle in the sand* and achievements, like *reach the other side*.

2. A deeper and less well recognized property of predicates is:

   Theorem 5. Validity (truth in all models) in a first order L is decidable if all lexical P_n's in L are P_1s. Decidability is lost if we add a single P_n for $n > 1$ (see Boolos and Jeffrey 1980 for proof).

So we see that adding two (or greater)-place predicates to a language increases its logical complexity. There is no longer a mechanical procedure which will tell us that an arbitrary sentence is true in all models if it is and that it isn't true in all models if it's not. And many issues of concern in linguistics center on P_ns for $n \geq 2$: the study of argument agreement with predicates, valency affecting operations (passive reduces arity, causatives and applicatives increase it), reflexives and reciprocals only arise with verbs that have two or more arguments, etc.

3. FOL has been very helpful to linguists in representing scope ambiguities. Note first that (7a, b) are logically distinct sentences in elementary arithmetic, (7a) being true and (7b) being false:

> (7)   a. $\forall x \exists y (x < y)$, *"For all x there is a y such that x is less than y"*
>
>   b. $\exists y \forall x (x < y)$, *"There is a y such that for all x, x is less than y"*

In (7a), the existential quantifier $\exists y$ is within the scope of the universal $\forall x$. So (7a) is true in a model iff no matter what element of the universe we let x denote, the formula $\exists y (x < y)$ is true, and that is so iff given the denotation of 'x' we can find a y in the universe such that y is greater than it. This of course is easy, just choose y to be $x + 1$. So for different choices of x we make different choices for y. But in (7b), we must choose a y first and then try to show that we can choose x freely always obtaining something less than y, but this cannot be. Choose x to be $y + 1$ for example.

Now linguists find a scope ambiguity in sentences such as (8a), which may be understood as in (8b) or (8c):

> (8)   a.   Some editor reads every manuscript
>
>   b.1.   There is at least one editor who reads all the manuscripts
>
>   b.2. $\exists y (\text{editor}(y) \mathbin{\&} \forall x (\text{manuscript}(x) \rightarrow y \text{ read } x))$
>
>   c.1.   For each manuscript there is an editor who reads it
>
>   c.2. $\forall x (\text{manuscript}(x) \rightarrow \exists y (\text{editor}(y) \mathbin{\&} y \text{ read } x))$

In (8b.1), the universal *every manuscript* is within the scope of the existential *some editor*, so on this reading (8a) is true iff there is a fixed editor who stands in the **read** relation to all the manuscripts. By contrast, the reading of (8a) expressed in (8c.1) is one in which it may be true if different editors read different manuscripts even though no one editor reads them all.

A last major contribution of FOL concerns variable binding. Compare (9a) from arithmetic with (10a) from English.

(9)   a. $\forall x(x \leq x^2)$

      b. $\forall x(x \leq y^2)$

(10)   a.   Every child loves his mother

      b.1. $\forall x(\text{child}(x) \rightarrow x \text{ love } (x\text{'s mother}))$

      b.2. $\forall x(\text{child}(x) \rightarrow x \text{ love } (y\text{'s mother}))$

In (9a, b) the superscript 2 is a one place function symbol. And our semantics for universal quantification says that (9a) is true iff no matter what number n we let x denote, $n \leq n^2$. The crucial point here being that we are relating each number n with its own square. But in (9b) the denotation of y^2 is given by context — whatever the assignment function says y denotes, all x are less than or equal to its square. It does not vary with the choice of x.

Now (10a) in English is ambiguous between readings expressed by b.1 and b.2. (10b.1) asserts a relation between each child and his/her own mother. But in (10b.2) we understand that the mother each child is related to doesn't vary with the child but is determined by the context of utterance. So again FOL provides a clear and precise way to semantically disambiguate natural language expressions.

Of note: several languages related to English present two different lexical items translating English 'his' in (10a), the one corresponding to the bound usage in (10b.1) the other to the unbound usage in (10b.2). Norwegian uses *sin* vs *hans*, Latin *suus* vs *eius*, Russian *svoi* vs *ego*.

Adequacy of FOL as a model of natural language semantics   Of course natural languages present many semantic properties FOL makes no attempt to represent: tense, imperatives, subjunctives, non-truth functional subordinate conjunctions like *because* (The truth of *John left early* because *the children were crying* can vary even if both the underlined sentences are true), etc.

More interestingly, how enlightening is FOL in representing the natural language phenomena (predicate-argument structure, quantification, binding) that linguists have drawn on it for? The semantic classification of predicates is a major topic of linguistic investigation, but not one we pursue here. We focus on the semantics of nominal quantifier phrases and, to a lesser extent, nominal modification by adjective phrases. Binding is independent of quantification. Here are some linguistically unnatural aspects of quantification in FOL that we modify and extend in this work.

First, compare (11a,b,c) in which we see that proper nouns (*Ted*) and quantified nominal expressions (*some teacher, each student*) are similar in that they seem to occur in the same positions satisfying argument requirements of the predicate. And importantly they are all syntactic constituents of the entire expression they occur in:

(11) a. Ted criticized some teacher

     b. Some teacher criticized Ted
$\exists x(\text{teacher}(x) \,\&\, \text{criticize}(x, \text{ted}))$

     c. Some teacher criticized each student
$\exists x(\text{teacher}(x) \,\&\, \forall y(\text{student}(y) \to \text{criticize}(x, y)))$

Linguists agree that these English sentences all consist of a "subject" nominal followed by a syntactically complex one place predicate (*criticized Ted/some teacher*, etc). In general syntactic constituents are meaningful, assigned a semantic interpretation. For *Ted* this might be (see later) an object in the universe of objects under discussion, per FOL. But quantified nominals, like *some/every student*, are not logical constituents and not assigned an interpretation in FOL. This is jarring in the extreme. In order to semantically interpret sentences like (11b,c) must we really tear apart syntactic constituents like *some student*? If so, why do we speak the way we do rather than in "logic-ese" per the FOL representation of sentences like (11b,c)?

Second the fact that proper nouns and quantified phrases have similar privileges of occurrence in (11) suggests that they should be treated in a semantically similar way. This is further supported by the fact that they may coordinate with each other — and typically the terms of a coordinate structure are meaningful in comparable ways. *Ted and some student were taking tickets when I arrived* is natural, but *Both Ted and laughed* doesn't make sense.

Third, and relatedly, in simple cases proper nouns and quantified nominals antecede ("bind") reflexive pronouns in the same way:

(12) a. Ted criticized himself (at the meeting).

     b. Some/Every worker criticized himself (at the meeting).

FOL handles (12b) naturally: $\exists x(\text{worker}(x) \,\&\, \text{criticize}(x, x))$ for example. But (12a) would be represented by something like Criticize(Ted,Ted), which is not so natural in English. *Ted criticized Ted* is awkward, and suggests that the speaker is referring to two different individuals named 'Ted'. We could always use $\exists x(x = \text{Ted} \,\&\, \text{criticize}(x, x))$, but this treats (12a) as the result of quantifying into a coordinate structure which seems foreign to the syntax of (12a).

It would not be hard to modify FOL so that an individual constant c also formed a variable binding operator (c, x), setting $(c, x)\varphi$ to be true iff φ is interpreted as 1 when the assignment sets x to denote $m(c)$. But our concern runs deeper here. The standard quantifiers $\forall x$ and $\exists x$ bind variables in the same way but have a different predicational meaning. We make this distinction formally in the interest of clarity. As we see later this turns out to be easy, with variable binding handled separately by a lambda operator and quantified nominals interpreted directly. But we anticipate.

Fourth, consider (13a,b). They have the same syntactic form in English, differing just by *some* vs *every*.

(13)   a.   Some poets daydream   a'. $\exists x(\text{poet}(x) \,\&\, \text{daydream}(x))$

       b.   All poets daydream   b'. $\forall x(\text{poet}(x) \to \text{daydream}(x))$

But their representations in FOL are significantly different. That they differ in quantifier symbol is natural as *all* and *some* differ in meaning. But the existential quantifier in (13a') quantifies into a conjunction of sentences whereas the universal quantifier in (13b') quantifies into a conditional sentence. Why should their main connectives be different? And there are, as we see later, many other quantifiers in English. Will they require different boolean connectives? (There are only 16 extensionally distinct two place boolean connectives but many more non-synonymous quantificational expressions.)

Fifth, and very significantly, English presents many quantifiers that are not definable in FOL. (We say explicitly what this means in Chapter 4.) Consider:

(14)   a.   <u>Most poets</u> daydream

       b.   <u>Forty percent of American teenagers</u> are overweight

       c.   There are <u>more cats than dogs</u> (in the garden)

       d.   <u>More women than men</u> signed the petition

The quantified nominals in (14a,b) refer to proportions of a set and as such are not first order definable, exceptions being the extremal cases like **100% of the = all**, etc. We take *most* in the sense of **more than half**; note too that (14a) has the same syntactic form as (13a,b) but has no first order representation at all! Equally the cardinal comparative quantifiers in (14c,d) lie beyond the first order boundary. We define proportionality and cardinal comparative quantifiers in later chapters.

Lastly, one reflection both pro and con. A merit of representing English semantics in first order is that we use a very limited set of primitives: n-place function and relation symbols, predicate-argument structures, boolean compounds and quantifications. Characterizing the many in terms of the few is a positive step in understanding.

On the other hand it would be discouraging to think that the wealth of structure in a natural language could be semantically characterized in a page or two. Good descriptive grammars of a natural languages run many hundreds of pages. So while the use of FOL has clearly proven insightful, that it is comprehensive is surely not the case. Indeed we study at some length in this work quantifier structures in English which are not expressible in first order. Finally, a brief word about Second Order Logic.

Second Order Logic is an enrichment of FOL in which for each n, we have variables ranging over the n-place relations on the universe. Thus we can form sentences such as (15):

(15) a. $\exists R \forall x \forall y (xRy \leftrightarrow yRx)$ "There exist symmetric relations"

b. $\forall P((P0 \ \& \ \forall n(Pn \rightarrow P(n+1))) \rightarrow \forall n Pn)$
"If P holds of 0 and it holds of $n+1$ whenever it holds of n then P holds of every natural number"

In (15b) the only second order variable that occurs ranges over subsets of the universe, possible P_1 denotations, so (15b) is a sentence in Monadic Second Order Logic (*Monadic* means we just quantify over one place predicates). A variety of natural quantifiers we consider are found in Monadic Second Order.

We also note that quantification over n-ary functions on the universe is properly second order, as seen in (16):

(16) For all F from E into E, if F is bijective then so is F^{-1}.

We note that in discussing first order expressive power later we focus on predicates rather than functions as they arise most naturally in our natural language discussion. Here we note that this apparently restricted focus does not really limit our (elementary) claims about expressive power since, while sometimes awkward, we can always mimic the behavior of an n-place function (one that maps n-ary sequences of elements of E to elements of E) using $n + 1$-ary relations ($n + 1$-ary predicate symbol denotations). To say that an $n + 1$-ary R determines a function we just require that if $(b_1, \ldots, b_n, x)$ and $(b_1, \ldots, b_n, y) \in R$ then $x = y$. This says, in effect, that R is a function.

We note that Second Order Logic, even Monadic Second Order, exceeds the Lindström barrier. Such Ls do have sound deductive systems but do not have complete ones, validity is not decidable, and they in general fail to be compact or have the interpolation property.

Appendix

Definition 1. SL is the closure of BF under AND, OR and NOT, where we set V to be BF $\cup$ {and, or, not,), (}. Then, SL $=_{\text{def}} \bigcap \{K \subseteq V^* \mid$ BF $\subseteq K$ and K is closed under AND, OR and NOT }. A subset K of V^* is closed under AND iff AND$(p, q) \in K$ whenever both p and q are. Analogously for OR and NOT.

An alternate definition of closure that is commonly used is:

Definition 2. Set SL$_0 =$ BF and SL$_{n+1} =$ SL$_n \cup \{$AND$(p, q) \mid p, q \in$ SL$_n\} \cup$ {OR$(p, q) \mid p, q \in$ SL$_n\} \cup \{$NOT$(p) \mid p \in$ SL$_n\}$.
Then SL $=_{\text{def}} \bigcup_n$ SL$_n$.

These two definitions provably yield the same set.

Chapter 3

Classifying Nominal Quantifiers in English

We are concerned first in this chapter to present a descriptively oriented overview of expressible types of nominal quantifier phrases in English. We supply some modest information concerning the distribution of these types in other languages. Our English inventory is largely drawn from Keenan (1996), Keenan and Stavi (1986), Keenan and Moss (1985), and Keenan and Westerståhl (1997). Our cross language claims are based largely on KP (Keenan and Paperno 2012) and PK (Paperno and Keenan 2017), which profited from Matthewson (2008), Suihkonen (2007), Suihkonen and Solovyev (2013), and Bach et al. (1995).

English presents an extensive and productively formed set of quantificational expressions. Linguists have tended to regard quantifiers as basically listable, primarily the existential and universal ones they inherited from elementary logic.

Second, we provide a compositional semantic interpretation for most of the quantifier types presented. It is important to see that the expressive power of natural language quantifiers goes well beyond the first order boundary and that nominal Quantified Phrases (QPs) can be directly interpreted — they need not (and often cannot) be torn apart and tied to variable binding operators ranging over the universe. Further, to directly interpret English QPs, we don't need more than the most elementary set theoretic operations. We provide an "interlude" below which spells out precisely what is meant by first order definability, a notion that is fundamental in previous studies of the logical expressive power of natural languages.

In later chapters we study the mathematical properties of the quantifiers of interest, supporting non-trivial generalizations that remain unstated and unsuspected in the absence of a mathematical means for characterizing the quantifiers. Here we also point out certain, perhaps unexpected, linguistic correlates of mathematically different classes of quantifiers. We begin:

Consider the following three sentences (Ss; which we also call P_0s, *zero place predicates*) which differ just by the choice of quantifier, (which we also call a *Determiner* or *Det*):

(1)   Some / All / Most poets daydream

Syntactically we treat these Ss as consisting of a (nominal) QP (Quantifier Phrase; we often say DP for Determiner Phrase) — *some poets*, *all poets*, *most poets* followed by a one place predicate (P_1), *daydream*. The P_1 may be complex: *daydream in the morning*, etc. Similarly, the nominal *poet(s)* may be complex: *romantic poets*, *romantic poets John likes*, etc. Shortly we consider many complex Dets. So the rough syntactic form of (1), ignoring tense, is given by:

(2)

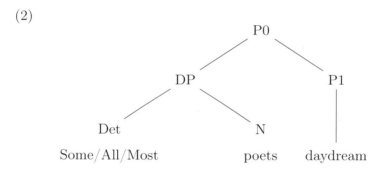

Semantically we *initially* adopt the primitives, E and $\{0,1\}$, of FOL. (Chapter 12 shows how to replace the universe E with a booleanly structured set P without loss of expressive power). P_0s (Ss) denote in $\{0,1\}$ and P_1s denote functions from E into $\{0,1\}$, as in FOL. We treat P_1s as subsets of E, isomorphic to the set of functions from E into $\{0,1\}$ as previously noted. So for p a P_1 denotation, it doesn't matter whether we say $p(b) = 1$ or $b \in p$, b an element of E. We shall refer to such sets (or functions) as *properties* (of elements of E) and we variously use variables $A, B, \ldots$ or $p, q, r, \ldots$ to range over properties.

An aside: treating P_1s equivalently as $P(E)$ or as $[E \to \{0,1\}]$ is our first instance of treating something in distinct but intertranslatable ways. This is common in mathematical work and has important advantages. Even if we can translate between two conceptualizations without loss of information, one approach may suggest questions and generalizations that the other does not. So we strongly advocate:

If you can't say something two ways you can't say it

As well, reformulating something already known in different terms can aid greatly in distinguishing the important properties of your objects of study from those that are artifacts of notation.   *End aside.*

We treat common noun phrases (*poet, romantic poet, ...*) as denoting properties. And we treat QPs (DPs) as functions from properties to truth values (of type (et,t) in Montagovian notation). Such functions are called *generalized quantifiers of type 1* (in Lindström 1966 notation). Just what function a DP of the form Det+N denotes depends on the denotations of its parts. The Ns can denote various properties. Our interest is the Det denotations. They are functions mapping properties to type 1 quantifiers. We say just what functions *some*, *all*, and *most* denote, using a very limited range of set theoretical notation:

(3)   For all subsets A, B of E
    a.   **some**$(A)(B) = 1$ iff $A \cap B \neq \varnothing$ (iff $|A \cap B| > 0$)
    b.   **all**$(A)(B) = 1$ iff $A \subseteq B$ (iff $|A - B| = 0$)
    c.   **most**$(A)(B) = 1$ iff $|A \cap B|/|A| > \frac{1}{2}$,   (iff $|A \cap B| > |A - B|$)
                       A finite & $0 < |A|$

Notation. $A \cap B$ (read A *intersect* B) is the set of objects common to A and B; $\varnothing$ is the empty set, the unique set with no members; $A - B$, A *minus* B (or the *complement* of B relative to A; now often written $A \setminus B$) is the set of objects in A which are not in B. Note that $A - B$ is empty if and only if (iff) everything in A is in B, that is, A is a subset of B, noted $A \subseteq B$. These set theoretic operations are the basic boolean operations on sets, to which we should add $A \cup B$, A *union* B, the set of things which are in A or B (or both). Also, for any set X, $|X|$ is the *cardinality* of X, roughly, the number of elements of X.

And at the risk of boring the reader, let us state that to *define* a function F from a set A, called the *domain* of F, to a set B, called the *codomain* of F, we must say for each $x \in A$ just what object in B F assigns to x. This object is noted $F(x)$, the *value* of F at x. More generally to define a function F you must say what its domain and codomain are and what value F takes at each x in its domain.

The domain of **some** (**all**, **most**) is $P(E)$, the set of subsets of E, and its codomain is the set of functions from $P(E)$ into $\{0, 1\}$. In general we write $[A \to B]$ for the set of functions from A into B, often (but not here) noted B^A. So in defining **some** (**all**, etc.) we must say for each $A \subseteq E$, what **some**(A) is. As an element of $[P(E) \to \{0,1\}]$ we define it by saying what its value is at each $B \subseteq E$. **some**(A) maps to 1 just those B which have a non-empty intersection with A. Functions of this type are of type (1,1), as they map two properties, successively, to a truth value. We will see higher types in what follows. For example, in *more students than teachers came to the lecture*, *more ... than ...* combines with two property denoting expressions to form a type 1 expression and so is of type ((1,1),1).

The quantifiers we defined in (3) are each representative of much larger classes. And they generalize in at least two directions.

Existential Quantifiers   Consider **some**. Its value at a pair A, B of properties is decided by looking at a single property, $A \cap B$. If that set is non-empty **some** maps A, B to true, otherwise to false. We shall call **some** intersective, and define (given arbitrary E, a point we will not repeat):

Definition 1. A function D of type (1,1) is *intersective* iff for all subsets A, B, X, Y of E, if $A \cap B = X \cap Y$ then $DAB = DXY$.

So an intersective D identifies pairs of properties with the same intersection. But **some** has an equivalent definition which suggests a stronger characterization. The value of **some**AB doesn't depend on which objects are in $A \cap B$, only on its cardinality, $|A \cap B|$.

Definition 2. D of type (1,1) is *cardinal* iff for all $A, B, X, Y \subseteq E$, if $|A \cap B| = |X \cap Y|$ then $DAB = DXY$.

All cardinal D are intersective. (Assume D is cardinal. Let $A \cap B = X \cap Y$, so trivially $|A \cap B| = |X \cap Y|$, whence $DAB = DXY$, so D is intersective.) We consider the more interesting converse shortly. We note that all languages in KP/PK can express low numerals and have various means of modifying numerical expressions. (Not all distinguish an indefinite article *a/an* from the numeral *one*; in WALS slightly less than 1/4 of the 473 languages sampled had an indefinite article or affix distinct from *one*. Historically in English *a/an* is a phonological reduction of *one*.)

(4)   a.   Bare numerals
           a/an, one, two, ..., a hundred, ..., a thousand, ...

      b.   Modified numerals
           exactly/precisely/only/just ten, more than/less than/fewer than ten, at most/at least ten, no more than ten, five or more

c. Vague modifiers
approximately/about/around a hundred, nearly/almost a dozen, a hundred give or take, practically/almost no

d. Bounding
between five and ten, a hundred plus or minus ten

e. No, hardly any, a dozen, a couple a'dozen, several, a score of, infinitely many, just finitely many, an even/odd number of

f. Boolean compounds
not more than ten, at least five but not more than ten, neither exactly ten nor exactly twenty, either about three or else more than fifty, not one, not a single

The vague cardinal Dets are of some interest, as it might seem that we can say little precise concerning *About 50 socks are in my drawer*. Does 46 count as about 50? If so then what about 45?, 44? So we may know exactly how many socks are in my drawer and not be clear whether there are about 50. But we can assert with some confidence that if the number of socks in my drawer is the same as the number of sparrows on my clothesline then *About 50 socks are in my drawer* and *About 50 sparrows are on my clothesline* have the same truth value — either both true or both false. Thus, *about 50* is cardinal.

We have not included the Dets in (5) though Barwise and Cooper (1981) and Lappin (1988) would include *many* and *few*.

(5)   many, few, too many/few, surprisingly many/few, (not) enough, more than enough, insufficiently many

These Dets focus on cardinality, but also contain a value judgment and per Keenan and Stavi (1986), fail to be extensional. Compare:

(6)   a. Not enough doctors attended the meeting

b. (More than) enough lawyers attended the meeting

Imagine a situation in which the doctors and the lawyers happen to be the same individuals (perhaps unbeknownst to everyone). And suppose that the meeting in (6) is of a medical association in which 250 doctors are required for a quorum but only one lawyer (to take the minutes). Just 200 individuals show up. We have enough lawyers then but not enough doctors. So Dets like *not enough* and *too many* do not depend for their value just on which individuals have the property the noun denotes. I limit myself here to extensional Dets, ones that do so depend. E.g. *every* is extensional: in the situation in (6) *Every doctor attended...* and *Every lawyer attended...* have the same truth value.

The linguist might balk at technical Dets like *infinitely many* where the truth conditions invoke a learned definition (due to Dedekind). Namely, a set X is infinite iff you can map it one to one onto a proper subset of itself. This definition is not part of our everyday competence. Still, if sets X and Y have the same cardinality then either both are infinite or both finite, so **just finitely many** and **infinitely many** are cardinal, whether they are "real" English or not.

Thus English has infinitely many cardinal Dets and so infinitely many intersective ones.

Query Are there intersective Dets in English which are not cardinal?

I have three candidates for an affirmative answer. None are simple. First, consider (7a) with *no... but Ted* interpreted as a Det per (7b).

(7)   a.   No student but Ted passed (the exam)

      b.   (**no... but Ted**)$(A)(B) = 1$ iff $A \cap B = \{$Ted$\}$

(7b) tells us that (7a) is true iff Ted is the only student who passed, which is correct. And **no... but Ted** is clearly intersective as its value at A, B is decided just by checking $A \cap B$. But it is not cardinal, it has to know that $A \cap B$ is $\{$Ted$\}$, not the equinumerous $\{$Sam$\}$.

A second candidate is interrogative *which?*. Contrast:

(8)   a.   How many students received scholarships?

      b.   Which students received scholarships?

Clearly *How many?* is cardinal, as *How many As are Bs?* is answered by giving the cardinality of $A \cap B$. But to answer *Which As are Bs?* we must identify the elements of $A \cap B$, not just its cardinality. These examples could be inconclusively discussed further. We forbear, noting only the obvious: they require extending our semantic analysis to questions, a major undertaking we have not done.

Our third example is of a different character and studied in a broader setting immediately below. Here we just briefly note that we can treat extensional adjectives like *bald* and *tall* as functions from properties to properties. Of interest here are those we call *absolute*, like *male* and *female*. A female artist is an artist who is a female individual, a male nurse a nurse who is a male individual, etc. (A tall nurse is a nurse, but need not be a tall individual.) Formally:

Definition 3. $F \in [P(E) \to P(E)]$ is *absolute* iff $\forall p \in P(E)$, $F(p) = p \cap F(E)$.

Now consider Dets of the form *more male than female*, using absolute adjectives. They are intersective but not cardinal:

(9)   For F, G absolute, (**more F than G**)$(p)(q) = 1$ iff $|F(p) \cap q| > |G(p) \cap q|$, iff $|p \cap F(E) \cap q| > |p \cap G(E) \cap q|$.

And clearly this last inequality remains true replacing p with s and q with t provided $p \cap q = s \cap t$. So *more male than female* is intersective, but it is not cardinal, as if we only know that $|p \cap q| = |s \cap t|$ we cannot infer that $|p \cap F(E) \cap q| > |p \cap G(E) \cap q|$. For counterexample let $p = \{a, b\}$, $q = \{b, c\}$, and $F(E) = \{a, b, c\}$, and $G(E) = \{x, y, z\}$. Then $|p \cap q \cap F(E)| = 1 > |p \cap q \cap G(E)| = 0$. Set $s = \{x, y\}$ and $t = \{y, z\}$. So $|p \cap q| = 1 = |s \cap t|$ but $|s \cap t \cap F(E)| = 0 \not> |s \cap t \cap G(E)| = 1$. So **more male than female** is not cardinal. Its value does not depend just on the cardinality of the intersection of its arguments p, q; it also needs to know the identity of male and female individuals.

A linguist might object to interpreting *more male than female* directly as a Det, favoring an approach which paraphrastically derives *more male than female poets* from *more male poets than female poets*. Such approaches increase the complexity of semantic interpretation, requiring interleaving interpretation between steps in syntactic generation or reconstruction of an underlying source from the observable surface form. Our direct interpretation approach is easy and natural, and avoids these syntactic complexities.

A superficially similar case that uses composition of Dets with adjectives occurs in "right node raising" (RNR) contexts: *No male and almost no female doctors object to that.* Generally, intersective Dets (**no**) composed with absolute adjectives yield intersective functions of type (1,1) which are not cardinal.

Let us turn now to the rich class of cardinal Dets in type ((1,1),1). Consider *more... than...* below which takes two property denoting arguments to yield a type 1 function.

(10)   (**more A than B**)$(X) = 1$ iff $|A \cap X| > |B \cap X|$.

The truth value assigned to X by (**more** A **than** B) is decided by the cardinality of the intersection of X with each of the Noun properties. These Dets have not been well studied (see Keenan 1987b) and some might doubt that they should be considered as arguments of predicates at all. In (11) we list some of the variety of such Dets, and (12) shows that they share distribution with other DPs.

(11)   (many) more ...than..., fewer...than..., (exactly / at least / nearly
        / almost) as many...as..., twice as many...as..., five times as
        many...as..., not more...than..., not very many more...than...,
        (almost) half again as many...as..., five more... than..., (about /
        approximately / nearly / almost) the same number of ...as...

The interpretation of such two place Dets is straightforward on the pattern of (10). We note further that cardinal comparison is not first order definable (Boolos 1981).

Such QPs occur as subjects (12a), but also as objects (12b), objects of prepositions (12c), as possessors (12d); they "raise to object" in (12e), passivize to subject (12f), and may occupy multiple argument positions of a given predicate simultaneously (12g,h).

(12)   a.   Half as many students as teachers attended the party

       b.   Ralph bought the same number of knives as forks

       c.   Ned solicits donations from as many unions as businesses

       d.   More men than women's applications were rejected

       e.   Sue believes more women than men to be qualified for the job

       f.   More women than men are believed to be qualified for the job

       g.   More men than women read fewer poems than plays

       h.   Twice as many women as men read as many poems as plays

Regarding boolean compounds, some Det_2s negate naturally enough, but conjunctions and disjunctions are at best awkward:

(13)   a.   Not more men than women were admitted

       b.   No fewer students than teachers attended the lecture

       c.?? At least as many as but not more, men than women attended

Below, we give a richer variety of Det_2s which combine with modifying adjectives to form Det_1s which "multiply out" to give the semantic effect of a Det_2, as we saw above. Compare:

(14)   a.   *More male than female* students were drafted

        b.   *More male* students *than female* students were drafted

        c.   *More male* doctors *than female* engineers were rejected

        d.   *Twice as many liberal as conservative* candidates won

        e.   *nearly/exactly/at least as many young as old* lawyers wept

So some Det_1s express comparison as well, and their adjectives may iterate: We interpret *female lawyer* as **female(lawyer)**, *young female lawyer* as **young(female(lawyer))**, etc.

Lastly here, note a somewhat surprising use of *or* among the nominal arguments of some Det_2s (see Keenan and Moss 1985 for some, limited, discussion. Here we only note the example). (15a,b) seem logically equivalent, with the *than* in (15a) behaving like *not* in: $\text{not}[p\text{ or }q] \equiv [\text{not } p \,\&\, \text{not } q]$.

(15)   a.   More young children than either adults or teenagers survived

        b.   More young children than adults survived and more young children than teenagers survived

The data in (16) highlight an inadvertent limitation on our use of type 1 DPs to date. Namely, they occur not only as subjects but quite generally also as objects of verbs and prepositions and as possessors. Several semanticists (Heim and Kratzer 1998) find this problematic, citing a "type mismatch" in the interpretation of (16b).

(16)   a.   *Some student / every student* offended John

        b.   John offended *some student / every student*

Semantically, the DPs in (16a) map a subset of E (those who offended John) to $\{0, 1\}$. But in (16b), *offended* is a P_2 and so denotes not a subset of E but a binary relation on E. The object DP does not have relations in its domain and so cannot assign a value to **offend**, "so" we cannot interpret *offended some / every student*.

In fact this "type mismatch" is a non-problem (Keenan 2016). DPs quite generally combine with P_{n+1}s to form P_ns, so they should be interpreted as functions from $n + 1$-ary relations to n-ary ones (0-ary ones denoting truth values 0,1). So their domain is the set of $n + 1$-ary relations over E.

Here is a solution: let F a function: $P(E) \to \{0, 1\}$. Add binary relations R to its domain, setting the value of F at each such R to the set defined by: $x \in F(R)$ iff $F(\{y \mid (xRy)\}) = 1$. So **offend some student** denotes a set of objects, the set of x which are such that **some student** holds of $\{y \mid x \text{ offend } y\}$. Thus it is true iff **student** has a non-empty intersection with that set. In general when R is an $n + 1$-ary relation, $F(R) =_{\text{def}} \{\langle a_1, \ldots, a_n \rangle \in E^n \mid F(\{b \mid \langle a_1, \ldots, a_n, b \rangle \in R\}) = 1\}$.

Note that the value of F at an $n+1$-ary relation is determined by its values at the unary relations, the subsets of E. So we can (continue to) define type 1 functions just by giving their values on the subsets of E, their values on other relations being determined as above. Keenan and Westerståhl (1997) treat DPs this way unproblematically. See Keenan (2016) for more discussion.

Universal Quantifiers   Our second example in (3) is *All poets daydream*. We see from (17) that the truth value of *all As are Bs* depends, a little surprisingly, on $A - B$, the As that are not Bs. If that set has an element then *All As are Bs* is false; otherwise true. A type (1,1) Det will be called *co-intersective* if it meets the invariance condition in (17). (18) gives some examples of Dets which meet this condition.

(17) D of type (1,1) is *co-intersective* iff for all $A, B, X, Y \subseteq E$,
        if $A - B = X - Y$ then $DAB = DXY$

(18)   all, every, each, all but two, all but at most ten, every... but Ted, nearly/almost/not quite all, not all/every, either all or nearly all

(19)   a.   (**all but n**)$(A)(B) = 1$ iff $|A - B| = n$

        b.   (**every** A **but Ted**)$(B) = 1$ iff $A - B = \{\textbf{Ted}\}$

Functions like **all but n** are co-cardinal and **every... but Ted** is co-intersective but not co-cardinal. The co-intersective Dets are less varied than the intersective ones, though the numeral in *all but n* guarantees that there are infinitely many.

A few smaller remarks about co-intersective Dets: First, one doesn't usually think of *all* in co-intersective terms though it does mean the same as the unnatural *all but zero*. Second, linguists have noticed that *all* and *each* differ in that the former has collective and mass uses and the latter is more rigorously distributive and limited to count nouns. The collective/distributive distinction, illustrated in (20a,b) is made in all the languages in KP/PK. Occasionally a mass term with a distributive quantifier forces a "kind" reading, (21b).

(20)   a.   All the monks met at matins in the morning (when else?)

   b.   *Each monk/*Each the monk met at matins. . .

(21)   a.   All tea contains caffeine

   b.   Each tea was exquisite (= Each type of, choice of, tea. . .)

Here we do not discriminate between *all*, *each*, and *every*. We note from KP/PK that all languages can express universal quantification, virtually always with a lexical item (though it doesn't always have the distribution of English *all* (*each*, *every*). Some languages have more than three lexical universal quantifiers.

   Third, *almost/nearly all* presupposes that the set quantified over is not too small. *Almost all students worry* is true in a model with 100 students, exactly 99 of whom worry. But if there are just two students and just one worries this is not clear, yet the number of students who don't worry is the same in both cases.

   The co-intersective functions are similar in some respects to the intersective ones — in both cases their value at a pair of properties is determined by a single property (given as a boolean function of the two in the pair). Further, as we show in the next chapter, these two sets of functions are booleanly isomorphic.

   But the two sets do differ in several ways. First, they are disjoint except for the unit function which maps all properties to 1, and the zero function which maps all properties to 0. Both these functions are intersective and also co-intersective. But otherwise, if there are some vegans among the students we can infer that some students are vegans (and some vegans are students) but we don't have the relevant information to infer that all students are vegan. Dually, given a student who isn't a vegan we know that *All students are vegans* is false, but we can't infer the truth value of *some students are vegans*.

   Second, all the Det$_2$s we gave, such as *more. . . than. . .*, were cardinal and thus intersective. There seem to be no Det$_2$s which are co-intersective. A possible exception to this claim might be the use of *every... and...* which, if treated as a Det, may take n nominal arguments all (finite) $n > 1$:

(22) a. Every man and woman jumped overboard $\equiv$ every man and every woman jumped overboard

     b. Every man, woman, and child jumped overboard $\equiv$ every man and every woman and every child jumped...

And three, empirically DPs built from intersective Dets occur very naturally in Existential There contexts whereas universal type DPs are much less natural there (in English), (23d):

(23) a. There are between five and ten students enrolled in the class

     b. Was there no one but John in the building then?

     c. Aren't there more women than men on the debate team?

     d. *Was there every student (but John) in the building?

A more prolific class of Dets is instantiated by *most* in (3).

Proportionality Quantifiers   These quantifiers have been little studied by linguists or logicians until recently. They lie beyond the first order boundary (see Barwise and Cooper 1981 for **most = more than half**, Westerståhl 1989, and especially Peters and Westerståhl 2006, Chapters 11–15, particularly Chapter 14.2:466), and so their entailment patterns have not been syntactically characterized. Yet they are quite productive in English. They serve semantically to stipulate the proportion of elements in a set that has a certain property and so we assume the set they quantify over is finite and non-empty.

(24) a. (*At least*) *Seven out of ten* sailors smoke Players

     b. (*Less than /at most*) *half the* students got an A on the exam

     c. *More than a third of* American teenagers are overweight

     d. *Not one* student *in ten* can answer that question

     e. *Just ten per cent of* government workers get raises every year

     f. *Between a quarter and a half of the* workers got a raise

It seems that proportionality Dets in English are syntactically somewhat complex. Still about half the languages in (KP/PK) have a lexical *half*.

Most is more typically complex (*la plupart de* 'the majority of'), but KP/PK cite a good half dozen monomorphemic *most*, as in English, Russian and Hebrew.

There appear to be *co-proportional* Dets, as in (25a), but it is logically equivalent to (25b):

(25)   a.   All but at most a tenth of the students passed

        b.   At least nine-tenths of the students passed

Like the intersective and co-intersective Dets proportionality ones form boolean compounds reasonably naturally:

(26)   a.   not more than half / three quarters

        b.   at least a third but not more than half

        c.   neither less than a quarter nor more than three quarters

        d.   between a third and a half

Also proportionality quantifiers overlap non-trivially with intersective and co-intersective ones. Of course the zero and unit functions are proportional, but also: *some ≡ more than 0%, no = (exactly) 0%, all ≡ 100%, not all ≡ less than 100%, some but not all ≡ more than 0% but less than 100%.*

And as with cardinal quantifiers, proportionality quantifiers also have two place variants:

(27)   a.   *Proportionately more* women *than* men signed the petition

        b.   *The same percentage of* boys *as* girls laugh at funny faces

        c.   *Ten per cent fewer* students *than* teachers objected to that

Interlude: Definability in First Order Logic   We have mentioned, without support or even explicit definition, that proportionality quantifiers as well as cardinal comparatives are not definable in FOL (First Order Logic). Here we make explicit just what this means.

But first, why do we care? There are several reasons. For one, FOL is very well understood, as we summarized briefly in Chapter 2. Thus it serves as a standard for logical expressive power. If the formal language in which we could give the semantic interpretation of English expressions was a first order language we would know that whenever an English sentence φ entailed an English sentence ψ then we could, in principle, find a (syntactic) proof from φ to ψ. And whenever some φ was valid (logically true) we could prove that too. On the other hand, if the semantic interpretation of English could be given just using SL (Sentential Logic) then there would be a mechanical procedure (truth tables) for proving φ if in fact φ was logically true and proving its negation if it wasn't.

But it turns out that English expressive power is way beyond that of SL in which the only truths and entailments are due to the meaning of the boolean connectives *and, or, not*, and those definable in terms of them. Lacking predicate-argument structure, SL cannot even show that (28a) entails (28b).

(28)   a.   All poets are vegetarian and Umberto is a poet

        b.   Umberto is a vegetarian

This entailment is demonstrable in FOL, so SL is not sufficient to express the entailment relation in English and FOL comes closer.

Early on in studies in generative grammar, linguists sometimes thought that English semantics could be given in a mildly strengthened version of FOL (Chomsky and Lasnik 1977, p. 429). A few scholars offered some subtle sentence types, e.g. branching quantification, whose naive semantic analysis went beyond first order (Barwise 1979; Hintikka 1973; Gabbay and Moravcsik 1974; Liu 1996) but many of the judgments of truth and entailment were not too convincing (Fauconnier 1975b).

Also (see below), a few quantifiers like *infinitely many* and *just finitely many* were easily shown to be not first order, but their mathematically technical character removed them from a position of central interest. One could maintain that English was "at heart" first order, acknowledging a few "mathematical" exceptions.

But now we know that a great many (indeed infinitely many) fairly ordinary quantifiers, such as the proportionality ones and the cardinal comparatives in (14) Chapter 2 are not first order definable. Let us illustrate now first some quantifiers, including some "non-standard" ones that are first order definable. Then we say what that means in general and then illustrate a few English quantifiers that are not definable in FOL.

Some first order definable quantifiers   Of course $\exists$ and $\forall$ are first order definable, trivially, as they are explicitly part of FOL (though only one of them need be taken as primitive. (We can define $\exists x$ in terms of $\forall x$ by: $\exists x \varphi \equiv \neg \forall x \neg \varphi$; and we can define the universal quantifier in terms of the existential by: $\forall x \varphi \equiv \neg \exists x \neg \varphi$.) In FOL, these quantifiers are understood to quantify over the entire universe of the model. Thus $\exists x\, cat(x)$ means "There is an x such that x is a cat", that is, "Some entity is a cat". And $\forall x \, \text{cat}(x)$ means "Every entity is a cat".

In ordinary English we normally accompany quantifiers with a CNP which specifies the domain over which we quantify (over which the variables range, sticking with the FOL terminology). Thus (29a) is an English sentence with the CNP *saint* denoting the domain of quantification (the restriction on the range of the variable). (29b) is the format we use for defining the denotation of *all*, and (29c) illustrates that our definition is expressible by a first order sentence, hence **all** is first order definable.

(29) a. All saints smile
      b. **all**$(p)(q) = 1$ iff $p \subseteq q$ $\qquad$ (iff $p - q = \varnothing$)
      c. **all**$(p)(q) = 1$ iff $\forall x(p(x) \to q(x))$ $\qquad$ ($\equiv \forall x(\neg p(x) \text{ or } q(x))$)

Similarly *at least two*, *all but one* and *exactly two* are first order:

(30) a. (**at least 2**)$(p)(q) = 1$ iff $\exists x \exists y(p(x) \& p(y) \& \neg(x = y) \& q(x) \& q(y))$

      b. (**all but one**)$(p)(q) = 1$ iff $\exists x(p(x) \& \neg q(x) \& (\forall y)((p(y) \& \neg q(y)) \to y = x))$

      c. (**exactly 2**)$(p)(q) = 1$ iff $\exists x \exists y(p(x) \& p(y) \& \neg(x = y) \& q(x) \& q(y) \& \forall z((p(z) \& q(z)) \to z = x \text{ or } z = y))$.

The numeral in (30a,b,c) can be replaced by any finite numeral preserving first order definability. Functions from pairs of properties, ones of type $(1,1)$, expressed by: *at least n*, *exactly n*, *more than n*, and *less than n* are first order definable as long as n is a finite numeral. This fails for cardinals like *infinitely many* and *just finitely many*.

In general, to show that a function F from pairs of properties to truth values is first order (definable) we must establish:

(31) For all universes E, all $p, q \subseteq E$, $F(p)(q) = 1$ iff φ, where φ is a first order sentence whose non-logical constants are limited to two one place predicate symbols (which may occur many times as may the boolean connectives, $=$, $\forall$, and $\exists$).

Commonly, proofs that one or another function F is not first order require some mathematical sophistication. Consider for example the argument that *infinitely many* is not first order.

(32)   (**infinitely many**)$(p)(q) = 1$ iff there is a proper subset s of $p \cap q$ such that there is a bijection from $p \cap q$ onto s.

Our defining sentence above is at least second order as it quantifies over sets and functions. But that does not imply that there is no first order sentence that would do the job. In fact there isn't, but how to show that? Here is a proof (which I give, as it illustrates the compactness property of FOL, one that appears to be of very little linguistic application on first reading).

Recall that a first order L is *compact*, meaning that whenever a set S of sentences entails a sentence ψ then some finite subset K of S also entails ψ. Now consider S and ψ below, where we write NumName for the set of names of natural numbers in English:

(33) $S = \{$At least n cats are on the mat $\mid n \in$ NumName$\}$
      $\psi =$ There are infinitely many cats on the mat

S is an infinite set of first order sentences (given that we can name infinitely many natural numbers). And S clearly entails ψ since for any number n we pick there are more than n cats on the mat. So if all the sentences in S are true then so is ψ. Now if *infinitely many* is first order, then by compactness some finite subset K of S entails ψ. Let K be any finite subset of S. Then there is a greatest natural number n such that some φ in S says that at least n cats are on the mat. That set K has a model in which there are exactly $n + 1$ cats on the mat. So K is true in a finite model, so K does not entail ψ. Contradiction. Hence *infinitely many* is not first order. (Hence *just finitely many* is not either, as (**just finitely many**)$(p)(q) \equiv$ ¬**infinitely many**)$(p)(q)$.   □

Now, as noted earlier, if the only English quantifiers that were not first order were technical ones like *infinitely many, just finitely many, uncountably many*, etc. we could pass these off as mathematical technicalities. But this is not the case. As we have previously supported with references, non-extremal proportionality quantifiers are not first order definable.

And equally cardinal comparison is not first order. We have not discussed this category in much detail, but our intuitions regarding truth conditions of sentences like (34) are clear and distinct:

(34)   a.   More men than women get drafted

       b.   Twice as many students as teachers attended the rally

Note that, most naturally, these quantifiers are of type $((1,1),1)$ — they map three properties to a truth value:

(35) $(\mathbf{more}\, p\, \mathbf{than}\, q)(s) = 1$ iff $|p \cap s| > |q \cap s|$

So to show first order definability of *more... than...* as above, we need a first order sentence φ whose non-logical constants are limited to three one place predicates and whose truth conditions are the same as those given in (35). And there aren't any. *End of interlude*

Definite Determiners We begin with a few examples. Those in (36a) are deictic, meaning their denotation is given as a function of properties of their utterance. Such dependencies are not studied here.

(36) a. my, this, those, your

 b. the (ten), John's (two), both John's, John's father's two
 the ten boys, both John's cats, John's father's two sisters

 c. the first/last/next...(that) we visited
 the first village (that) we visited

 d. the first/last/next...to set foot on the moon
 the first person to set foot on the moon

 e. the youngest / friendliest...(that) we met

The is a definite article, and based on the 566 language sample in WALS, about half the world's languages have a definite article distinct from a demonstrative (like *that*). All languages have at least one such demonstrative. In several cases a definite article derives historically from a reduced demonstrative (Latin *ille* $\Rightarrow$ French *le* 'the'). A first pass at a semantics for *the n* is:

(37) $(\mathbf{the\ n})(A)(B) = 1$ iff $|A| = n$ and $\mathbf{all}(A)(B) = 1$

We treat *the one* simply as *the* when followed by a singular noun. The information that $|A| = n$ in (37) is in some sense presupposed. The natural way to query whether *The two boys are asleep* is *Are the two boys asleep?* — which does not query whether there are two boys but only whether they are asleep. We do not enter into presupposition here though *both* and *neither* (*both pupils, neither pupil*) presuppose that $|\mathbf{pupil}| = 2$. See partitives below.
  Of interest are (36c,d,e) in which the postnominal relative clauses are understood to form a semantic unit with the prenominal ordinals or superlatives. The first village we visited is the village we visited before we visited any other village. The first woman to set foot on the moon need not, in

any absolute sense, be the first woman (?Eve?). And the youngest student we met may not be the youngest student in the model, but just one who is younger than all the other students we met. English has other constructions which interpret as a constituent discontinuous prenominal and postnominal material. *That's an easy theorem to state but a difficult one to prove. A smarter linguist than Sue would be hard to find.* A case of direct interest to us is what we may call *Free Choice Determiners*:

(38) a. Help yourself to whatever drinks you find in the fridge

      b. *Help yourself to whatever drinks

We are tempted to read (38a) as ... *whatever you find in the fridge drinks*, which isn't really English. But we need the relative clause somewhere, as (38b) shows.

Definite Dets differ from the other Dets so far considered in not forming boolean compounds naturally: *not the two boys, *not John's cat, *the two but not the three cats, .*?neither these nor those cats.* Bare possessors do coordinate, though adding numerals is marginal: *neither John's nor Mary's articles were accepted*; *??Neither John's two nor Mary's three articles were accepted*.

The DPs, like *the two boys, John's doctor*, etc. built with definite Dets are also called *definite*. Possessors of definite DPs themselves are definite Dets. Definite DPs, even more than co-intersective and proportionality ones, do not easily occur in Existential There contexts in English, (39a), but they are natural in partitives, (40a):

(39) a. *?Aren't there all / most students in the class?

      b.   *Aren't there the two students in the class?

(40) a.   Two of the ten, two of John's ten, two of my ten (children)

      b. *?Two of no boys / ten boys / most boys / not all boys

Partitive Dets: Det of Det$_{def}$ Here, following Keenan and Stavi (1986) we treat *two of the ten* as a complex Det in *two of the ten cats*. More usually, linguists assume *the ten cats* is a DP but do not interpret the whole partitive expression. In English, *both* and *neither* as Dets seem to be lexical partitives: **both/neither ≡ each/not one of the two**. KP cite Japanese and Finnish as having a lexical *Which of the two?* Gitksan has a lexical *some of the* (PK).

(41)   a.   some / (almost) all / almost/a third of the ten / John's pets

         b.   not one / not any / not more than ten of John's students

         c.   [most of Ann's but hardly any of Ted's] articles

         d. *most of no boys / *not all of most students

         e. *the four of the ten / *the four of John's ten dogs

The examples in (41a,b,c) show that the (co-)intersective and proportional Dets occur fairly freely in the first Det position in partitives. (41d) supports that the second Det should be definite and (41e) shows that the partitive construction does not iterate, that is, the first Det is not itself naturally a partitive.

Chapter 4

Generalized Quantifiers and Logical Expressive Power

This chapter focuses on *generalized quantifiers* of type 1. They are expressed by DPs which combine with P_{n+1}s to form P_ns.

We exhibit many pairs of logically equivalent sentences (Ss) built from the same verbs but differing by the choice of argument DP. We seek, and find, logical forms for such Ss from which their logical equivalence is provable. The argument DPs include non-first order definable proportionality ones (*more than half the students*) and cardinal comparatives (*more students than teachers*). FOL, recall, is essentially the strongest logic in which the (semantic) entailment relation is syntactically characterizable. The entailment patterns build on three types of semantic negation. Philosophers may note that these patterns (massively) generalize the classical Aristotlean square of opposition (Appendix 2). Mathematicians may note that they illustrate the Klein four group acting on generalized quantifiers (Appendix 3). We begin:

1. Some queries
Query 1 The (a,b) pairs below are logically equivalent (true in the same models). What is their "logical form" which predicts this?

(1)   a.   At least two thirds of the students answered no question at all.

      b.   At most a third of the students answered at least one question.

(2)   a.   Every student but John read at least as many plays as poems.

      b.   No student but John read fewer plays than poems.

(3)   a.   Almost all stockbrokers read at least one financial paper at breakfast.

      b.   Hardly any stockbrokers read no financial paper at breakfast.

(1) uses a proportionality quantifier, (2) a cardinal comparative one, and (3) an inherently vague quantifier, which seems not definable at all. The truth of (3a,b) may fail to be determined even knowing how many stockbrokers there are and how many read a financial paper at breakfast.

Query 2 The first two (a,b) pairs below are logically equivalent but just differ in that the main predicate of one is the negation of that of the other. The third (a,b) pair seem to differ in the same way, but logical equivalence fails. What is the logical form of these QPs which accounts for these facts?

(4) a. Between a third and two thirds of the students passed

   b. Between a third and two thirds of the students didn't pass

(5) a. Either every student but Ed or no student but Ed got an A

   b. Either every student but Ed or no student but Ed didn't get an A

(6) a. Between a third and three quarters of the students passed

   b. Between a third and three quarters of the students didn't pass

Query 3 The (7a,b)–(9a,b) pairs, now using three arguments, are logically equivalent. Again, what is their logical form?

(7) a. Each counselor told both John and Bill at least three stories

   b. No counselor told either John or Bill fewer than three stories

(8) a. Not every witness told every detective two or more lies

   b. At least one witness told some detective fewer than two lies

(9) a. All but one witness told more than half the jurors at least one lie

   b. Just one witness told at least half the jurors no lie at all

Query 4 In (7)–(9) the middle DPs are not identical. But in (10) and (11) they are. What property do these DPs have which justifies this? Why does the pair in (12) fail logical equivalence?

(10) a. Both counselors told John at least three stories

    b. Neither counselor told John fewer than three stories

(11)   a.   All but one judge awarded himself at least one prize

     b.   Exactly one judge awarded himself no prize at all

(12)   a.   Both counselors told every camper at least three stories

     b. $\not\equiv$ Neither counselor told every camper fewer than three stories

2. Types of Semantic "Negation"
2.1 Boolean Complement

Observe that if (13a) is true then (13b) is false, and conversely:

(13)   a.   Ted laughs at my jokes

     b.   Ted doesn't laugh at my jokes

We will say that such Ss stand in the *boolean complement* relation. This relation is expressed by negation (*not/n't*) in this and many other examples. Ss denote in a set with a boolean structure, which supports *meet* ($\wedge$), *join* ($\vee$) and *complement* ($\neg$) operations often denoted by *and*, *or*, and *not* respectively (regardless of whether Ss denote truth values, "propositions", functions from possible worlds to truth values, etc.).

P_1s, one place predicates, like *laughs at my jokes* also denote in a boolean structure and support a natural interpretation for *and*, *or*, and *not/n't*. *Doesn't daydream* holds of John iff *daydreams* fails of John. So too with DPs: *not all poets* holds of *daydream* iff *all poets* fails of *daydream*. Our logical constituency is shown in (14):

(14)   a.   [All poets][daydream]   a'. $F(p)$
     b.   [Not [all poets]][daydream]   b'. $(\neg F)(p)$
     c.   It is not so that [all poets daydream]   c'. $\neg(F(p))$
     d.   [All poets][don't daydream]   d'. $F(\neg p)$

For F, G functions from P_1 to S ($= P_0$) denotations (more generally from P_{n+1} denotations to P_n denotations, see **Appendix 1**) we have:

(15) $(F \wedge G)(p) = F(p) \wedge G(p);$
       $(F \vee G)(p) = F(p) \vee G(p);$
       $(\neg F)(p) = \neg(F(p))$

(16) For d a DP, *not d* generally denotes its boolean complement, as in (a). But there are other ways of expressing complement, (b):

    a. not every student; not all poets; not a (single) boy; not more than ten cats; not both John and Bill; not more than half the dogs; not more boys than girls

    b.

X	COMPLEMENT Y
More than ten boys	At most ten boys
At least half the plays	Less than half the plays
Either John or Bill	Neither John nor Bill
Some student	No student
At least two boys	Less than two boys
One or more girls	No girls
(At least) as many poems as plays	Fewer poems than plays
(At least) as many poems as plays	More plays than poems
At least ten students	Fewer than ten students
Exactly half the boys	Either less or more than half the boys

(17) The P_0 pairs below are boolean complements, as are their subject DPs:

    a. More than half the students passed

    a′. At most half the students passed

    b. Either Sue or Ann came early

    b′. Neither Sue nor Ann came early

(18) For F of type 1 mapping P_{n+1} denotations to P_n ones (**Appendix 1**), the value of F at a binary relation R, a possible P_2 denotation, is given by: $x \in F(R)$ iff $F(\{y \mid (xRy)\}) = 1$. So John is in the set denoted by *offend every linguist*, (**every linguist**)(**offend**), iff (**every linguist**) holds of the set of things that John offended, which is correct (see Keenan 2016).

(19) Facts re boolean complement, $\neg$:

    1. All natural languages can express it (almost always with overt morphemes, sometimes discontinuously, but see Pederson 1993 re Old Tamil).

2. $\neg$ is bijective, self inverting: $\neg\neg F = F$, and symmetric: $F = \neg G$ iff $G = \neg F$.

3. $F \neq \neg F$, ever. (I.e. boolean complement $\neg$ has no fixed points).

2. Post-Complement For F a GQ, define $F\neg$, read "F post-complement" by: $(F\neg)(p) = F(\neg p)$. So **not every student** is the postcomplement ("inner negation") of **some student** since (20a,b) are logically equivalent:

(20) a. Not every student passed.

      b. Some student didn't pass.

(21)   X   POSTCOMPLEMENT   Y

X	Y
every student	no student
both Ann and Sue	neither Ann nor Sue
every student but John	no student but John
more than 90% of the girls	less than 10% of the girls
(exactly) six outta ten sailors	(exactly) four outta ten sailors
at most a third of the boys	at least two thirds of the boys
all but a tenth of John's dogs	a tenth of John's dogs
exactly half the students	exactly half the students
not every student	some student
all but finitely many sentences	just finitely many sentences
both judges	neither judge
less than one boy in ten	more than nine out of ten boys
each of the ten boys	not one / none of the ten boys
almost all stockbrokers	hardly any stockbrokers
a majority of the students	a minority of the students
more than seven outta ten poets	less than seven outta ten poets

(22) Facts re boolean postcomplement

  1. Like boolean complement it is self inverting: $F\neg\neg = F$, symmetric: $F = G\neg$ iff $G = F\neg$, and bijective.
  2. Unlike boolean complement:

      a. It has fixed points: for some F, $F = F\neg$ (examples shortly).
      b. Not all boolean algebras exhibit postcomplements. For $(F\neg)(p) = F(\neg p)$ to make sense p itself must lie a boolean structure, so that the domain of F is a boolean lattice.

3. Empirically, I think, the DPs built by cardinal Det$_2$s, *more students than teachers* do not present natural expressions of post-complement.

4. **Conjecture** No natural language has a monomorphemic expression interpreted as the postcomplement operation.

(23)   **Observations explained**
The "logical form" of [$_S$ DP$_1$ V DP$_2$] is: $F(G(R))$, where DP$_1$ and DP$_2$ (which may be proper nouns) denote the GQs F and G respectively, and the transitive verb denotes the binary relation R. And **Query 1** claims that for certain GQs F, G, F', G':

(24) $F(G(R)) = F'(G'(R))$, all binary relations R.

So given F and G what choices may we make for F' and G' so as to make (24) true? The answer is surprisingly constrained. First we eliminate two degenerate cases:

Definition 1. A GQ H is *trivial* iff H is constant (maps all p to 1, or all p to 0).

Example, *either at least six or fewer than six individuals* is trivial, its denotation maps all p to 1. The result of replacing *or* by *and* maps all p to 0. And obviously if F is trivial then $F(G(R) = F(G'(R))$, all GQs G'. Now we explain **Query 1**:

Theorem 1 (Facing Negations (weak version)**).** For F, G, F', G' any non-trivial GQs, R any binary relation
$$F(G(R)) = (F\neg)((\neg G)(R).$$

Proof.
$$
\begin{aligned}
(F\neg)((\neg G)(R)) &= (F\neg)(\neg(G)(R)) & \text{Def complement} \\
&= F(\neg(\neg(G(R))) & \text{Def postcomp} \\
&= F(G(R)) & \neg \text{ is self inverting} \quad \square
\end{aligned}
$$

Theorem 1 explains the (a,b) equivalences in (1)–(3) as the (a) sentence has the form F(G(R)) and the (b) one $(F\neg)((\neg G)(R))$:

Theorem 2 (Facing Negations (Keenan 1993)**).** For F, G, F', G' any non-trivial GQs, R any binary relation
$$F(G(R)) = F'(G'(R)) \text{ iff } (F = F' \ \& \ G = G') \text{ or } (F' = F\neg \ \& \ G' = \neg G).$$

Resolving Query 2 The subject of (4a,b), and (5a,b) denote F that are their own postcomplement: $F = F\neg$, so these pairs are of the same form as those in (1)–(3).

$$(\textbf{Between 1/3 and 2/3 of the } A)(B) =$$
$$(\textbf{Between 1/3 and 2/3 of the } A)(\neg B) =$$
$$((\textbf{Between 1/3 and 2/3 of the } A)\neg)(B).$$

The subject DP in (6a,b) does not denote an F which is its own postcomp so by **Theorem 2** the two Ss are not logically equivalent.

Notation. We may simplify our statements by writing $(F \circ G)(R)$ for $F(G(R))$ where $\circ$ just indicates function composition (Note that $(F \circ G)$ maps P_{n+2}s to P_ns). Facing Negations now says that $(F \circ G) = (F\neg \circ \neg G)$. Note too that function composition is associative: $(F \circ G) \circ H = F \circ (G \circ H)$, so we can write simply $F \circ G \circ H$. We note:

(25) a. $\neg(F \circ G) = (\neg F) \circ G$

b. $(F \circ G)\neg = F \circ (G\neg)$

c. $(F \circ G)$ is non-trivial iff both F and G are.

Resolving Query 3 applies the complement and the postcomplement operations simultaneously to a given GQ. We note that we may omit parentheses: $(\neg(F\neg))(p) = \neg((F\neg)(p)) = \neg(F(\neg p)) = (\neg F)(\neg p) = ((\neg F)\neg)(p)$. $\neg F\neg$ is often noted F^{d} and read "F dual". A classical example of duals is **every** and **some**:

(26) Every pupil passed $\equiv$ No pupil didn't pass $\equiv$ It is not so that some/any pupil didn't pass.

As is apparent, checking that $F = \neg G\neg$ may involve Ss with two negations, often awkward to process. So to check that $F = \neg G\neg$ it is often easier to check the equivalent statement: $\neg F = G\neg$. (Note: $F = \neg G\neg$ iff $\neg F = \neg(\neg G\neg)$, iff $\neg F = G\neg$.)

(27)   X   DUAL   Y
          Every student   some student
          Both John and Bill   either John or Bill
          At most 70% of the students   less than 30% of the students
          All but at most two students   more than two students
          More than 9 of the 15 students   at least 6 of the 15 students
          Less than one baby in three   at most two out of three babies
          Both students   at least one of the two students
          Not both of the students   neither student

Accounting for Queries 3 and 4   The GQ sequences in (7)–(9) are of the (logical) form $F \circ G \circ H$ and $F\neg \circ \neg G\neg \circ \neg H$, and so identical by Facing Negations, so the Ss are logically equivalent. (10a,b) are also of this form as *John* denotes an individual which is self dual (below). (12a,b) are not equivalent since *every camper* does not denote a self dual function. What about the middle DP *himself* in (11)? Technically I should not have included *himself* as it does not map P_1s to P_0s. But it does map P_2s to P_1s. And it commutes with boolean complement and so is self-dual and we predict the equivalence of (11a,b).

(28) $x \in \mathbf{self}(R)$ iff $(x, x) \in R$, so $x \in \mathbf{self}(\neg R)$ iff $(x, x) \in \neg R$, iff $(x, x) \notin R$, iff $x \notin \mathbf{self}(R)$, iff $x \in \neg(\mathbf{self}(R))$.

Some properties of duals

  1. Like comp and postcomp duality is self inverting: $F^{dd} = F$, symmetric: $F = G^d$ iff $G = F^d$, and bijective.
  2. Like postcomplement they may have fixed points.
  3. F is self dual, $F = F^d$, iff F commutes with complement: $\neg F = F\neg$.

(29)   a.   Singular proper nouns denote self dual GQs:
            John isn't Greek $\equiv$ It is not the case that John is Greek.

       b.   Given $b \in D$, the domain of discourse, I_b is$_{\mathrm{def}}$ that map sending a subset A of D to 1 iff $b \in A$. The I_bs are called *individuals*. Provably they preserve complements, and so are self dual (see Zimmermann 1993). Singular proper nouns denote individuals.

NB Facing Negations applies to $n > 3$ place predicates (**Appendix 4**), though just how naturally is open (see Malagasy, Kinyarwanda, Turkish, Japanese):

(30) $F \circ G \circ H \circ I = F\neg \circ \neg G \circ H\neg \circ \neg I$

    a. Both bailiffs let 3 or more defendants pay each judge no bribe.

    b. Neither bailiff let fewer than 3 defendants pay no judge a bribe.

(31) $F\neg \circ \neg G\neg \circ \neg H\neg \circ \neg I$

    a. Each coach let both Ed and Al pay more than 2 referees no fee.

    b. No coach let either Ed or Al pay all but at most 2 referees a fee.

A Linguistic Generalization: the Possessive Construction Preserves Complement, Post-Complement and hence Duals Thus in (32) below the possessors of the left and right hand members are complements of each other, and so are the entire DPs. In (33) the possessors and the entire DPs are postcomplements of each other, and in (34) duality extends from possessors to the entire DP.

(32)   COMP

    a. Some student's advisor   a′. No student's advisor

    b. More than one child's mother   b′. At most one child's mother

(33)   POSTCOMP

    a. Both defendants' lawyers   a′. Neither defendants' lawyers

    b. Exactly half John's advisors   b′. Exactly half John's advisors

(34)   DUAL

    a. Every child's mother   a′. Some child's mother

    b. At most 2/3 of Ed's dogs   b′. Less than a third of Ed's dogs

(35) $F \circ G = F\neg \circ \neg G$

    a. Both defendant's lawyers raised at most three objections.

    b. Neither defendant's lawyers raised more than three objections.

(36) $F \circ G \circ H = F\neg \circ \neg G\neg \circ \neg H$

   a.   At least $2/3$ of the judges awarded every contestant's sponsor at least one prize.

   b.   Less than $1/3$ of the judges awarded any contestant's sponsor no prize at all.

Two Open Queries

Q1 Writing xR for $\{y \mid xRy\}$, the set of y that x stands in the relation R to, the functions F we have studied satisfy: if $aR = bS$ then $F(R)(a) = F(S)(b)$. So whether $F(R)$ holds of some x just depends the set of things x stands in the relation R to. But not so for *anaphors*: if John criticized just the individuals Bill praised it does NOT follow that *John criticized himself* and *Bill praised himself* have the same truth value. Ditto for the object DPs in *John criticized every student but himself*, *No one likes to work with anyone smarter than himself*, etc. Proper anaphors G must satisfy only: if $aR = aS$ then $G(R)(a) = G(S)(a)$. So our query is:

   When does $F \circ G = F' \circ G'$ when G, G' may be anaphors?

Q2 (37a,b) are logically equivalent. What is the generalization that covers these cases? Provably (Keenan 1992) we can represent the LF of these Ss (unpleasantly) by a function mapping P_{n+2}s directly to P_ns, but there are no two functions F, G of type P_{n+1} to P_n such that the interpretation of (37a) $= F(G(\mathbf{answer}))$.

(37)   a.   All the students answered the same questions (on the exam).

   b.   No two students answered different questions (on the exam).

Appendix 1

Definition 1. We define functions F from P_{n+1}s to P_ns as follows: Given a domain E, the set of n-ary relations over E is $P(E^n)$, where n ranges over the natural numbers. The domain of F then is $\bigcup_n(P(E^{n+1}))$, its codomain is $\bigcup_n(P(E^n))$, and for each $n + 1$-ary R, $F(R)$ is that n-ary relation given by: $(x_1, \ldots, x_n) \in F(R)$ iff $F(\{a \mid (x_1, \ldots, x_n, a) \in R\}) = 1$. No conditions are placed on F at unary relations (subsets of E), and the value of F at $n > 1$-ary relations is determined by its values at the unary relations.

Appendix 2

Traditional square of opposition　　　　Any Generalized Quantifier F

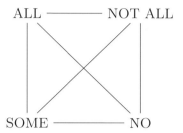
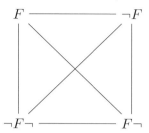

For candidate "squares" in other categories see Löbner (1986).

Appendix 3　　The Klein four group (noted **V** for *vierergruppe*)

$$
\begin{array}{c|cccc}
  & e & a & b & c \\
\hline
e & e & a & b & c \\
a & a & e & c & b \\
b & b & c & e & a \\
c & c & b & a & e \\
\end{array}
$$

Here each entry represents $x \cdot y$, where x is from the left column, y the top row. e is the identity element, $e \cdot x = x \cdot e = x$; $\cdot$ is commutative: $x \cdot y = y \cdot x$ (so **V** is Abelian) and the product of any two distinct non-identity elements is the third non-identity element. Our group consists of four functions from GQs to GQs: $\mathrm{id}(F) = F$, $\mathrm{comp}(F) = \neg F$, $\mathrm{postcomp}(F) = F\neg$, and $\mathrm{dual}(F) = F^{\mathrm{d}}$. The map φ sending e to id, the identity map from GQs to GQs, a to comp, b to postcomp and c to dual, is an isomorphism: $\varphi(x \cdot y) = \varphi(x) \circ \varphi(y)$.

Appendix 4

Theorem 3 (Facing Negations Generalized). Let $\mathbf{F} = (F_1, \ldots, F_{n+1})$ and $\mathbf{G} = (G_1, \ldots, G_{n+1})$ be sequences of non-trivial GQs of the type considered above. Then $\mathbf{F} = \mathbf{G}$ iff (a) or (b) holds:

(a) $(F_1, \ldots, F_n) = (G_1, \ldots, G_n)$ and $F_{n+1} = G_{n+1}$.
(b) $(F_1, \ldots, F_n) = (G_1, \ldots, G_n)\neg$ and $F_{n+1} = \neg G_{n+1}$.

Theorem 3 is quite surprising: the F_i in **F** can be *any* functions (of the appropriate type, not just "logical" ones (= ones invariant under permutations of the domain)). Note that while Dets such as *every, not more than half*, etc. denote invariant functions in their type, *every student*, etc. do not, due to the denotational variability of *student*. So you might think starting with a "random" F_n we could compensate for its idiosyncracy by using later counter-idiosyncratic F_i. But no, **Facing Negations** is the only non-trivial pattern.

Chapter 5

The Midpoint Theorems

Here we discuss briefly, two entailment paradigms naturally involving proportionality quantifiers (Keenan 2005, 2008). I had originally wondered if these paradigms might not be characteristic of proportionality quantifiers, but Peters and Westerståhl (2006) generalize insightfully, the crucial paradigm showing that it applies to a more general class of quantifiers, though initially it is awkward to find properly non-proportional examples that are naturally expressible. We provide information concerning the boolean structure of midpoint quantifiers which enable us to find a few others.

For the first paradigm, note that the (a,b) members of each pair below are logically equivalent.

(1) a. Exactly half the students passed the exam

b. Exactly half the students didn't pass the exam

(2) a. Between a third and two thirds of the students passed

b. Between a third and two thirds of the students didn't pass

From the perspective of standard logic this pattern is slightly surprising. For $Q = \forall$ or $\exists$ it never happens that $Qx\varphi \equiv Qx\neg\varphi$. It is also mildly surprising from a linguistic perspective. We don't normally expect a predicate and its negation to hold of the same subject. Indeed replacing *two thirds* in (2a,b) with any of the denumerably many other non-synonymous fractions, say *three quarters*, results in pairs that are not logically equivalent.

The subject DPs in (1) and (2) denote type 1 functions F that are their own postcomplements: $F = F\neg$. We shall follow Westerståhl (2012) in calling such quantifiers *midpoints*. So our query now becomes one of characterizing how to form expressions that denote midpoints (fixed points of the postcomplement function). A first generalization of the pattern in (1) and (2) is given in Keenan (2008).

One might have thought that the (1a,b) case was just an idiosyncracy of (*exactly*) *half*. But the (2a,b) pair generalizes the pattern. And it is not hard to see that the relevant choices of fractions are ones symmetrically distributed around the midpoint between 0 and 1 (equivalently 0% and 100%). Thus (Keenan 2008):

Theorem 1. For p, q fractions with $0 \leq p \leq q \leq 1$ and $p + q = 1$,

    a. **More than p and less than q** is a midpoint, and
    b. **At least p and not more than q** is a midpoint
    c. **Between p and q** is a midpoint

Of course p and q can be given as percentages: **between 40 and 60 percent (of)** as in *Between forty and sixty per cent of American teenagers are (are not) overweight*. In general *between p and q* can be understood strictly (open interval) as in (a) above, or weakly (closed interval) as in (b).

To discover further midpoints the following theorem is helpful:

Theorem 2. The postcomplement function from type (1) functions to type (1) functions is a boolean isomorphism (Keenan 2008).

Proof. $((F \wedge G)\neg)(p) = (F \wedge G)(\neg p)(F(\neg p) \wedge G(\neg p)) = (F\neg(p) \wedge G\neg(p))$, hence $(F \wedge G)\neg = (F\neg \wedge G\neg)$, so postcomplement respects $\wedge$. Also $(\neg F)\neg = \neg(F\neg)$, so postcomplement respects complement.   □

Theorem 2 enables us to show Theorem 3. Its easy corollary below is already noted in Westerståhl (2012):

Theorem 3. For all type 1 F, $(F \wedge F\neg)$ and $(F \vee F\neg)$ are midpoints.

Proof. $(F \vee F\neg)\neg = (F\neg \vee F\neg\neg) = (F\neg \vee F) = (F \vee F\neg)$; dually for $\wedge$.   □

Corollary. F is a midpoint iff for some G, $F = G \vee G\neg$ (or $= G \wedge G\neg$).

Theorem 4 expresses a proper generalization of (2a,b). Note that exactly $n\% =$ at least $n\% \wedge \neg$more than $n\%$. And we infer that the members of the pairs below are logically equivalent:

(3)   a.   At least three out of ten but not more than seven out of ten students will (will not) get scholarships next year.

        b.   Some but not all of the men laughed (didn't laugh) at that joke.

        b′.   Some of the men but not all of them laughed (didn't laugh) ...

c. Either no student but John or else every student but John got (didn't get) an A on the exam.

d. Either all of the boys or else none of them passed (didn't pass).

In each case in (3) the conjuncts are postcomplements, as are the disjuncts, so Theorem 3 applies. Another theorem helpful in identifying midpoints is:

Theorem 4. The set of midpoints is a complete sublattice of type 1. So the set of DPs which denote midpoints is closed under conjunction, disjunction and negation.

(4) a. Either none of the students or else all of them (don't) like jazz.

b. Either less than a third or else more than two thirds of the students are (are not) vegetarians.

The subject in (4a) is the complement of *some of the students but not all of them*, a midpoint. In (4b) the subject is the complement of *between a third and two thirds of the students* which is a midpoint, thus its complement is also by Theorem 4.

  Are midpoints always proportional? That is, are DPs which denote midpoints always built from proportionality Dets? The answer is no, yet the most natural examples seem to be. Even (3b), *some but not all*, is a boolean compound of proportional Dets since **some** = **more than 0% of** and **all** = **100% of**. Similarly the Dets in (4a) are proportional, but do not seem to be at first glance. But Theorem 4 leads us to notice some midpoints of a strictly non-proportional nature, e.g. (5a).

  Since a complete sublattice of a ca BL must itself be atomic it is natural to wonder what the atoms are and how we may express them, if at all. Now the DPs in (5) denote atoms in the lattice of midpoints and the Dets in (5a,b) are not proportionality ones:

(5) a. Either everyone but John or else no one but John

b. Either just the students or just the non-students

c. All poets and no non-poets or no poets and all non-poets

The direct way of expressing the first conjunct in (5c) uses the same boolean apparatus we use later in the Free Generator Theorem. Recall that for each $b \in E$, I_b is that type (1) function mapping a property p to 1 iff $b \in p$ (iff $p(b) = 1$). It is the *individual* generated by b. And using I as a variable ranging over individuals we have:

(6)   All p and no non-p $= \bigwedge\{I \mid I(p) = 1\} \wedge \neg \bigvee\{I \mid I(\neg p) = 1\}$

(7)   No p and all non-p $= \neg \bigvee\{I \mid I(p) = 1\} \wedge \bigwedge\{I \mid I(\neg p) = 1\}$

Clearly (6) holds just of p and (7) just of $\neg p$, so their disjunction holds just of p and $\neg p$. This makes them midpoints, and since they hold of something they are not zero, and since they hold of just one pair, $(p, \neg p)$ they are as small as possible satisfying the criterion for being midpoints, so they are atoms. In general an atom in the midpoint lattice is given by:

(8)   Either every p and no non-p or every non-p and no p

Also we see that (9) presents examples of DPs built from non-proportional partitive Dets:

(9)   a.   Exactly five of the ten students passed (didn't pass).

      b.   Either exactly one or else all but one of the students passed.

We note that replacing *five* in (9a) by *half*, a proportional Det, does not change its truth conditions. In general Dets of the forms in (10) build midpoint DPs:

(10)   a.   exactly n of the $2n$

      b.   exactly n or else all but n

We close this chapter with a brief look at a different entailment paradigm, (11), one that is again natural with proportional Dets.

(11)   More than two thirds of the students passed the exam.
      At least one third of the students are athletes.
      ───
      Therefore, at least one student is an athlete and passed the exam.

Keenan (2005) originally hoped that paradigms like (11) might be characteristic of proportionality Dets and relate them to better known classical quantifiers, like **some**. It does indeed do the latter, but not the former. Peters and Westerståhl (2006, p. 191) prove a proper generalization of (11) which covers more than proportionality Dets. To state the generalization we first define:

Definition 1. A type (1) function F is (*monotone*) *increasing* iff for all properties q, q' if $q \subseteq q'$ and $F(q) = 1$ then $F(q') = 1$. A function D of type

(1,1) is *right increasing* (increasing on its right argument) iff for all properties p, $D(p)$ is increasing. Dually, D is right decreasing iff Dpq and $q' \subseteq q$ implies Dpq'.

Recall also that a type (1,1) Det function D is conservative iff $D(p)(q) = D(p)(p \cap q)$, all properties p, q. Then:

(12) **Westerståhl's generalization**
For D conservative, [1] and [2] below are equivalent:

[1] D is right increasing
[2] $D(p)(q) \wedge D^{\mathrm{d}}(p)(s) \Rightarrow \mathbf{some}(p)(q \cap s)$

Our previous examples, e.g. (11), illustrate (12), but again it is hard to find natural examples built from Dets that are not proportional. Here is one, using partitive Dets:

(13)   More than five of the nine students are athletes.
       At least four of the nine students are vegans.
       —————————————————————————————
       Therefore some student is both an athlete and a vegan.

A slightly different, nearly trivial, non-proportional example is:

(14)   Both students are athletes.
       At least one of the two students is vegan.
       —————————————————————————————
       Therefore some student is both an athlete and a vegan.

The theorem in (12) has a kind of dual, (16). We give the proof, as it is not available in the literature (but it is basically just the dual of that in Peters and Westerståhl). Note that (12) claims there is non-empty overlap between two sets, whereas (16) claims there is a non-empty gap. First an illustrative entailment paradigm from English.

(15)   a.   Less than 30% of American teenagers are literate.

       b.   At most 70% of American teenagers are athletes.

            Ergo, at least one American teenager is neither literate nor an athlete.

(16)   For D conservative, [1] and [2] below are equivalent:

       [1] D is right decreasing

[2] $D(p)(q) \wedge D^{\mathrm{d}}(p)(s) \Rightarrow \mathbf{some}(p)\neg(q \cup s)$

Proof. [1] $\Rightarrow$ [2]: Given Dpq and $D^{\mathrm{d}}ps$. Assume, leading to a contradiction, that $p \cap \neg q \cap \neg s = \varnothing$. Then $p \cap \neg s \subseteq q$, so by the right decreasingness of D, $Dp, p \cap \neg s = 1 = Dp, \neg s$ by the conservativity of D, $= (Dp\neg)(s)$, so $\neg Dp\neg, s = D^{\mathrm{d}}ps = 0$, contradicting the assumption. So $p \cap \neg q \cap \neg s = p \cap \neg(q \cup s) \neq \varnothing$.

[2] $\Rightarrow$ [1]: Let $Dpq = 1$ with $q' \subseteq q$. Show $Dpq' = 1$.

Assume for contradiction that $Dpq' = 0$. Then $(Dp\neg)(\neg q') = 0$, so $\neg Dp\neg(\neg q') = 1 = D^{\mathrm{d}}p(\neg q')$, ergo $Dpq \& D^{\mathrm{d}}p(\neg q')$ so $\mathbf{some}(p)\neg(q \cup \neg q') = 1$, so $p \cap \neg q \cap q' \neq \varnothing$, so $\neg q \cap q' \neq \varnothing$, so $q' \not\subseteq q$, contradiction. Thus $Dpq' = 1$, so D is right decreasing. $\square$

And again the natural examples of this entailment paradigm seem to be ones with proportionality quantifiers. But, as with (13) and (14) we do find a few non-proportional examples:

(17)   Less than four of the nine students are vegan.
         At most five of the nine students are athletes.
         ⎯⎯⎯⎯⎯⎯⎯⎯⎯⎯⎯⎯⎯⎯⎯⎯⎯⎯⎯⎯⎯⎯⎯⎯⎯⎯⎯
         Ergo, some student is neither a vegan nor an athlete.

(18)   Neither student is vegan.
         At most one of the two students is an athlete.
         ⎯⎯⎯⎯⎯⎯⎯⎯⎯⎯⎯⎯⎯⎯⎯⎯⎯⎯⎯⎯⎯⎯⎯⎯⎯⎯⎯
         Ergo, at least one student is neither a vegan nor an athlete.

Chapter 6

Three Linguistic Generalizations

We ended Chapter 4 with a generalization concerning the relation between classical syntactic structure, possessive DPs, and their boolean semantic properties. Here, we refine and extend that generalization and then present two further generalizations. In all cases, the generalizations illustrate the utility of boolean concepts (but we are not remotely suggesting that there are no other useful and enlightening semantic concepts).

Our examples are intended not simply to illustrate the descriptive utility of characterizing linguistic phenomena in boolean terms, but to substantiate a much deeper point (one that is picked up again in a more creative way in Chapter 12). Namely: *To explain things we are trying to understand we should characterize them in terms of notions we already understand.* We count as already understood basic boolean notions, as this mathematical area is well studied and well understood. We discuss the epistemological issues somewhat more in Chapter 12. Here we begin with illustrative examples that will serve as background later.

In studying possessive DPs like *Mary's algebra teacher* or *every student's doctor*, noted here DP_{poss}, we refer to *Mary* and *every student* as Possessor DPs. They combine with *'s* to form possessive Dets, noted Det_{poss}, which combine with CNPs to form possessive DPs. These in turn may further combine with *'s* to form possessive Dets, which combine with further CNPs to form yet more complex DP_{poss}, etc. *Ted's football coach, Ted's football coach's employer, Ted's football coach's employer's union, Ted's football coach's employer's union's headquarters, ...*

So, iterating the steps which build a possessive DP seems natural enough. As with cases we will see later, the derived expressions get more unwieldy as they get longer, but there seems to be no natural cut off point. In this respect these derivational operations resemble simple mathematical functions. E.g. define the squaring function (SQ, usually noted superscript ^{2}) from the natural numbers $\mathbb{N}$ to $\mathbb{N}$ by: $SQ(n) = n \cdot n$. Then $SQ(SQ(SQ(3))) = SQ(SQ(9)) =$

SQ(81) = 6561. If we wanted to prevent the squaring function from applying to its own output we would have to change the codomain of SQ so that the set of SQ(n) had an empty intersection with the domain of SQ. There is nothing wrong with defining functions whose values do not lie in their domain, but to do so we have to impose conditions on their (co)domains explicitly to make this happen.

Many morphological derivational functions in English are of this sort — e.g. ones that change category. E.g. let LY be that function that applies to adjectives to yield adverbs: LY(*slow*) = *slowly*, LY(*careful*) = *carefully*, etc. But this function does not iterate: LY(LY(*slow*)) = LY(*slowly*) = **slowlyly*. (More precisely LY(*slowly*) doesn't make sense, since *slowly* is not in the domain of LY.)

But many syntactic derivational processes do iterate (Keenan and Moss 2016, Chapter 1): *He knows a doctor, He knows a doctor who knows a doctor, He knows a doctor who knows a doctor who knows a doctor, ...; Which thief took that painting?, Which detective knows which thief took that painting?, Which administrator knows which detective knows which thief took that painting?, ...*

But even in syntax, many operations change category and do not iterate: we might derive the nominal *John's laughing* from the sentence *John laughed*, but that derivation step does not naturally reapply: (**John's's laughinging annoyed me, *?John's laughing's annoying me*).

These observations actually help us understand one syntactic role of grammatical categories. They allow us to regulate recursion: if we limit a given derivational operation to expressions of a certain category, then an operation which derives an expression of a different category will not iterate. More generally one important role of grammatical categories is to provide and constrain domains of derivational functions.

A note: **Natural language recursion and infinity** For generative grammarians, recursion "glitters". It is often regarded as characteristic of human languages as distinct from animal signaling systems, which appear bounded with the number of structurally distinct (see Chapter 12) expression types being listable.

Here are two judgments about recursion which may remove some of its glitter without doubting its existence — *He knows a doctor who knows a doctor who...* is clearly an instance of recursion. First, recursion seems to me the **default** case for syntactic (as opposed to morphological) derivation. For a sequence of syntactic derivational operations not to iterate, we need to stipulate conditions on the domains and codomains of the generating functions;

if we say nothing, the derivation may iterate *ad infinitum*. And second, work within generative grammar assumes recursion but rarely studies it explicitly. The topics in textbooks on recursion theory (Boolos et al. 2007; Epstein and Carnielli 1989), e.g. the hierarchy of primitive recursive, recursive (= computable), recursively enumerable, etc. functions are not of concern to most linguists and not part of a usual PhD course program.

Linguists (and others) sometimes think that allowing or requiring the set of natural language expressions to be infinite increases the complexity of grammatical analysis. But quite the opposite is the case. Unmarked recursion results in infinitely many expressions (more than n for each natural number n). But if we required that e.g. no English expression have more than 10,000 words/morphemes/... our statement of the grammar would be much more complex, comporting various conditions limiting iteration of different interacting derivational processes so as to not cross the 10,000 boundary. In detail, this would be nightmarish and colossally uninteresting. So

Infinity Simplifies.

Worth noting is that the simplifying effect of the infinite shows up semantically. In a first order L the set of valid sentences (those true in all models) is recursively enumerable (but not computable). So if some φ is in that set, that fact is provable (though you may not be clever enough to concoct a proof). However, the set of sentences φ true in all *finite* models is not even recursively enumerable. So for an arbitrary φ in that set we are not guaranteed that the set of proofs of sentences in L contains one ending in φ.

End of note.

Monotonicity and Possessive DPs   We return now to our semantic generalization regarding possessive DPs. Chapter 4 illustrates that if two possessor DPs are boolean complements, such as *some student* and *no student*, then the DP$_{poss}$ built by combining them with the same CNP are also boolean complements: *Some student's bicycle was stolen* and *No student's bicycle was stolen* must have different truth values. Similarly if the two possessor DPs are postcomplements or duals so are the corresponding DP$_{poss}$. We now further note:

(1)   Let Y be a DP$_{poss}$ with possessor DP X. Then,

    a.   if X is increasing so is Y

    b.   if X is decreasing so is Y

    c.   if X is non-monotonic (neither increasing nor decreasing) so is Y

Thus *most opera singer's agents* is increasing because *most opera singers* is (If most opera singers read Goethe and everyone who reads Goethe is German then most opera singers are German). Similarly *at most two opera singer's agents* is decreasing because *at most two opera singers* is. And *exactly two opera singers' agents* is non-monotonic because *exactly two opera singers* is.

Monotonicity and Negative Polarity Items   Monotonicity properties have been of considerable interest to linguists as they interact in at least partially predictable ways with *negative polarity items* (NPIs). We define:

Definition 1. For f a function from a partially ordered set $(B, \leq)$ to a partially ordered set $(B', \leq')$,

    a. f *preserves the order* iff for all $x, y \in B$, $x \leq y \Rightarrow f(x) \leq' f(y)$

    b. f *reverses the order* iff for all $x, y \in B$, $x \leq y \Rightarrow f(y) \leq' f(x)$

    c. For *preserves (reverses) the order*, linguists may say *preserves polarity* and *reverses polarity* respectively.

NPIs were introduced by Klima (1964) as a listable number of usually small words and fixed phrases whose presence was conditioned by being under the scope of negation. We illustrate here with the NPIs *any* and *ever*:

(2)   a. Billy didn't see *any* birds on the walk

      b. Rachel hasn't *ever* been to Pinsk

(3)   a. *Billy saw *any* birds on the walk

      b. *Rachel has *ever* been to Pinsk

The sentences in (2) contain the negation *n't* and the presence of *any* and *ever* is licensed. Eliminating the negation as in (3) results in ungrammaticality, though replacing *any* by *some* in (3a) restores it, as does replacing *ever* by *occasionally* in (3b).

Later Ladusaw (1983), see also Fauconnier (1975a) who considers scale reversal for broader pragmatic scales, noticed that other "negative" expressions besides classical negation *n't* and *not* also license NPIs:

(4)   a. No / Fewer than five children saw *any* birds on the walk.

      b. At most two students here have *ever* been to Pinsk.

(5) a. *Some / *Five children saw *any* birds on the walk.

  b. *More than two students here have *ever* been to Pinsk.

These data raise two interesting questions, the first empirical, the second more conceptual:

(6) a. Which subject DPs license NPIs in the predicate, as illustrated in (4)?

  b. What semantic commonality is there between the licensing DPs in (4) and the negators *n't* and *not*?

Our second question presupposes that there is some semantic commonality between classical negation and the licensing DPs in (4) — intuitively they are all "negative", but just what does that mean?

  What we need is some "negative" property that relates classical negation *n't/not*, which seems to be an "adverb", and Dets such as *fewer than five, no, not more than ten*, etc. which are quite different from adverbs. But in all cases they take their denotations in a set with a boolean order $\leq$, and the NPI licensors have in common that the functions they denote reverse the order, per the Ladusaw-Fauconnier Generalization:

Ladusaw-Fauconnier Generalization

NPIs occur within the arguments of expressions
denoting decreasing functions (but not increasing ones).

Let us spell out that the relevant expressions here are indeed decreasing in this sense. First consider *n't/not*, which we take to denote the boolean complement function in whatever lattice/algebra the expression it combines with denotes in. First we note:

Theorem 1. In any boolean lattice $(B, \leq)$, for all $x, y \in B$, $x \leq y$ iff $\neg y \leq \neg x$.

  For example if B is $P(E)$, the set of subsets of a non-empty set E, and A, B are subsets of E, then $A \subseteq B \Rightarrow \neg B \subseteq \neg A$ (The converse also holds, but we do not need that here). If everything in A is in B then the set of things not in B is a subset of those not in A. Equally if p, q are sentence denotations, $p \leq q$ iff $\neg q \leq \neg p$. That is, if p implies q and q is false, then p is false, that is, not q implies not p. Equally, in any model in which everyone who is walking is talking, that is, **walk** $\leq$ **talk** in that model, then everyone who isn't talking isn't walking, that is, $\neg$**talk** $\leq \neg$**walk** in that model. Thus in any boolean lattice, the complement function reverses polarity.

Now DPs map properties p, elements of $P(E)$, to the truth value lattice $\{0,1\}$ in which for all $x, y \in \{0,1\}$, $x \leq y$ iff $x \to y$, that is, $x = 1$ implies $y = 1$.

So for F, G functions from $P(E)$ into $\{0,1\}$, $F \leq G$ iff for all $p \in P(E)$, $F(p) \leq G(p)$. ($F(p)$ and $G(p)$ are truth values so we know what $\leq$ means when applied to them). So we see that e.g. **John and some student** $\leq$ **some student**, since if the former holds of some property p then so does the latter. And we observe that **some student** preserves polarity: if $p \subseteq q$, i.e. all p's are q's, then if some student is a p then some student is a q. But **no student** reverses polarity: if all p's are q's and no student is a q then no student is a p; that is, $p \subseteq q \Rightarrow [(\textbf{no student})(q) = 1 \Rightarrow (\textbf{no student})(p) = 1]$. So *no student* is decreasing. Similarly one sees that *fewer than five, at most five, not more than five*, etc. build decreasing DPs.

On the importance of negative polarity items   Capturing the distribution of NPIs may seem like a minor issue in terms of designing a grammar to generate just the expressions competent speakers/signers of English regard as well formed, that is, as English. But as not infrequently happens, small issues, when pursued carefully, lead to much larger ones, as is the case here.

The bigger issue concerns the autonomy of syntax. Generative grammarians regard the presence of a language, and hence an internalized grammar, as biologically based and distinctively human. The idea is that we humans are biologically endowed with a core syntactic mechanism, one with many parameters that take values given by experience (English, Basque, Malagasy, ... are not identical). Of course we semantically interpret our innately given syntax in rational, utilitarian ways, but the syntax itself is not motivated by such interpretative concerns, it is what we are given to start with.

An analogy may help here. Imagine two groups of people on islands too far apart to shout to each other, but close enough to see flags of different colors and shapes. The two groups have flags of the same shapes and colors. Their colors and shapes have no inherent meaning, but to communicate — to say hello, trade jokes, and eventually prove theorems they start to wave their arbitrary flags in regular ways to establish signs for sunrise and sunset (and eventually Ode to a Nightingale and the Pythagorean theorem), etc.

Now, negative polarity items are an issue in this scenario as it appears that to know whether certain phonologically possible expressions are grammatical or not, it seems that we must know whether the expressions denote monotone decreasing functions or not. So our syntactic judgments of well-formedness would depend on semantic interpretation. This appears to be in violation of the view that our innate grammar is purely syntactic in nature, like the flags of random shapes and colors. It is as though our island dwellers had to know to begin with that a red triangular flag meant sunset.

A merit of the order reversing approach to NPIs is that it yields fairly good predictions concerning the monotonicity of booleanly complex Dets and DPs and hence of their potential for licensing NPIs. In DPs of the form Det+CNP it is the Det that determines whether the entire DP is increasing, decreasing or neither. E.g. if *no teacher* is decreasing so is *no student, no student in the class*, etc. More generally:

(7) **Fact**: For p, q CNPs and d a Det, $d + p$ and $d + q$ have the same monotonicity: both increasing, both decreasing, or both neither.

In passing, we note that for DPs of the form Det+CNP, certain properties of the DP are determined by the Det and others by the CNP. And as we have just seen, whether the DP licenses NPIs in these cases is decided by the Det: $no + p$ licenses NPIs, $some + p$ does not. But whether Det+CNP satisfies the selectional restrictions of a predicate is determined by the CNP: *%more than two ceilings laughed at my joke* is anomalous, so is *%every ceiling laughed at my joke*. But *more than two children laughed...* and *every child laughed...* are natural. Now observe the following easy theorem.

Theorem 2. For F, G generalized quantifiers (maps from $P(E) \rightarrow \{0, 1\}$),

   a. If F and G have the same monotonicity value (both increasing, both decreasing, both neither) then $F \wedge G$ and $F \vee G$ have that same value.
   b. F is increasing iff $\neg F$ is decreasing

Thus we predict the well formedness of the following sentences:

(8) a. Not more than two children saw *any* birds on the walk.

   b. No students and not more than two teachers here have *ever* been to Pinsk.

   c. None of the students and at most two of the teachers had *ever* seen *any* pregnant kangaroos.

   d. Either no student or at most one (student) saw *any* birds on the walk.

Note also that the two place boolean connective *neither... nor...* is something like a two place complement: it builds a decreasing function from two increasing ones, so we correctly predict:

(9)   Neither John nor Mary saw *any* birds on the walk.

In sum our response to the leading questions in (6) is that the subject DPs that license NPIs are those that denote decreasing functions, and secondly, the licensing DPs have in common with classical negation *n't/not* that they denote decreasing functions (with negation actually meeting a stronger condition). As well certain theorems concerning the preservation of monotonicity allows us to predict that various complex DPs license NPIs.

We should note that the range of syntactic environments that license NPIs extends well beyond classical negation plus decreasing DP subjects. Our purpose here has been primarily to illustrate the nature of a properly semantic generalization. How far we can extend this approach is a matter of research.

Our approach does contrast however with a more "linguistic" one, as presented for example in Linebarger (1987). She draws attention to several syntactic and pragmatic contexts besides those we have considered which license NPIs. On her analysis NPI licensors are "lexically marked with a contextual feature ... of immediate adjacency to negation".

I find this approach conceptually unsatisfactory in that we do not have independent access to these "contextual features" — indeed I am not even sure what to look for. They are attached to constituents of objects in "LF", but LF, whose members look like expressions in a language, is not defined. In contrast when we claim that *fewer than five students* is decreasing whereas *more than five* is not we argue from direct judgments of entailment. So reversal of the boolean $\leq$ is something we can independently support.

Cardinality DPs and Existential There Constructions   Since Milsark (1977) there is an extensive linguistic literature on which DPs occur naturally in Existential There (ET) contexts in English (see ter Meulen and Reuland 1987). (10) illustrates some examples pro and con with the relevant DPs underlined:

(10)   a.   There are *more than five students* in the class.

   b. *There are *the ten students* / *John's ten students* in the class.

In fact there is excellent agreement that DPs built from cardinal Dets as previously defined do occur naturally in ET contexts, as in (10a). That definite DPs, as in (10b), do not naturally occur there is somewhat tempered by "list context" (Rando and Napoli 1978) sentences, (11a), and superlatives (Fauconnier 1975b), (11b).

(11)   a.   – How do I get to UCLA from here?
             Well, there's always *the bus*, but it doesn't run very often.

       b.   There isn't *the slightest chance* that he will succeed.

Anna Szabolcsi (p.c.) points out to me that interpretative variation is less if interrogative or negative ET contexts are used. **Well, there isn't always the bus, but...* And in any event, the generalizations we support here concern DPs that do naturally occur in ET contexts.

We saw earlier that cardinality Dets are a special case of intersective ones, so we may wonder whether DPs built from non-cardinal intersective Dets also occur in ET contexts. They seem to.

(12)   a.   Was there *no one but Willie* in the building?

       b.   Just *which students* were there are the party anyway?

       c.   Weren't there *more male than female students* at the party?

Now recall DPs built from two place cardinal Dets such as *more... than...*, as in *more students than teachers* and *half again as many students as teachers*, etc. These DPs have two arguments and are cardinal in that whether they hold of a predicate property s depends on the cardinality of both $s \cap p$ and $s \cap q$, where p and q are the denotations of the two CNPs the Det$_2$ combines with. E.g.

(13)   (**more** p **than** q)$(s) = 1$ iff $|p \cap s| > |q \cap s|$

And as expected DPs built from these Dets occur quite naturally in ET contexts (Keenan 1987a), a fact unaccounted for on most linguistic treatments of ET sentences (McNally 2011, 2016).

(14)   a.   Aren't there *more women than men* in your class?

       b.   There aren't *twice as many men as women* in the class.

       c.   There weren't *exactly as many men as women* at the dinner.

Our last generalization regarding ET contexts seems innocuous:

(15)   ET contexts accept boolean compounds of D(P)s they accept.

(16) a. Aren't there are a couple of cats and a dog in the bedroom?

　　　b. There are at least two but not more than five cats on the mat

　　　c. There are no boys and not more than five girls in the class

　　　d. Aren't there some Norwegians or a few Swedes in the class?

Chapter 7

Mathematical Analysis of English Quantifiers

Boolean preliminaries Here we formulate some mathematical generalizations about the classes of quantifiers presented earlier. For this we need some boolean background. For linguistic naturalness we present boolean structure as relational lattices, interdefinable with boolean algebras.

Definition 1. A *lattice* is a pair $(L, \leq)$ where L is a non-empty set and $\leq$ is a partial order relation on L — reflexive: $x \leq x$; antisymmetric: $x \leq y \,\&\, y \leq x \Rightarrow x = y$, and transitive: $(x \leq y \,\&\, y \leq z) \Rightarrow x \leq z, x, y, z \in L$.

Further, for all $x, y \in L$, the set $\{x, y\}$ has a *greatest lower bound* (glb) and a *least upper bound* (lub), where, for $K \subseteq L$, a *lower bound* (lb) for K is an element z, not necessarily in K, such that (s.t.) $z \leq k$, all $k \in K$; z is a greatest lower bound (glb) iff z is a lb for K and $w \leq z$, all lbs w for K. If K has a glb it is unique and noted $\bigwedge K$. We usually write $x \wedge y$ (read: x *meet* y) as $\bigwedge\{x, y\}$.

Dually w is an *upper bound* (ub) for K iff for all $k \in K$, $k \leq w$. An ub w for K is *least* iff $w \leq v$, all ubs v for K. $\bigvee K$ is the unique lub for K if it has a lub. Write $x \vee y$ (read: x *join* y) as $\bigvee\{x, y\}$.

Definition 2. A *lattice* is *complete* iff all $K \subseteq L$ have a glb and a lub.

We note that if all K have a glb then all K have a lub, and conversely.

All the lattices we treat in this work are complete. Note that the definition of lattice only requires all $\{x, y\}$ to have a glb and a lub. It says nothing about the empty set or any infinite subsets.

Definition 3. A *boolean* lattice (BL) is a lattice $(B, \leq)$ that is bounded, distributive and complemented:

1. *Bounded*: B itself has a glb called the *zero*, noted 0, and a lub, called the *unit*, noted 1, and

2. *Distributivity*: $x \wedge (y \vee z) = (x \wedge y) \vee (x \wedge z)$ and $x \vee (y \wedge z) = (x \vee y) \wedge (x \vee z)$

3. *Complemented*: $\forall x \exists y$ such that $x \wedge y = 0$ and $x \vee y = 1$. For each x this y is provably unique and noted $\neg x$ (read *complement x*).

A prototypical example of a BL (boolean lattice) is a pair $(P(A), \subseteq)$, where $P(A)$ is the power set of a non-empty set A and $\subseteq$ is the subset relation on the elements of $P(A)$. Clearly $\subseteq$ is reflexive, antisymmetric and transitive. Given $X, Y \subseteq A$, $X \cap Y$ is the glb for $\{X, Y\}$ and $X \cup Y$ its lub. A and $\varnothing$ are the unit and zero elements, so $(P(A), \subseteq)$ is bounded. $X \cap (Y \cup Z) = (X \cap Y) \cup (X \cap Z)$ and dually, $X \cup (Y \cap Z) = (X \cup Y) \cap (X \cup Z)$ so distributivity holds. And finally $(P(A), \subseteq)$ is *complemented*: For all X there is a Y, namely $Y = A - X$, such that $X \cap Y = \varnothing$, and $X \cup Y = A$.

Also power set boolean lattices are complete: Let J be a set s.t. for each $j \in J$, U_j is a subset of A. Then $\{U_j \mid j \in J\}$ has a glb, namely the set of x that lie in all the U_j. This set is noted either $\bigcap \{U_j \mid j \in J\}$ or $\bigcap_{j \in J} U_j$. Similarly $\bigcup \{U_j \mid j \in J\}$ or $\bigcup_{j \in J} U_j$ denotes its lub.

We note that basically boolean lattices and boolean algebras are the same structures, they differ with regard to what is taken as primitive in their definitions (Grätzer 1998, p. 6). The relational lattice approach takes the $\leq$ relation as primitive and defines $\wedge$, $\vee$, $\neg$, 0, 1 in terms of it. The algebra approach takes these functions as basic and defines $\leq$ in terms of them (e.g. $x \leq y$ iff $x \wedge y = x$).

A second fundamental example of a BL is the set $\{0, 1\}$ of truth values. Usually it is given algebraically, with $\wedge$, $\vee$, and $\neg$ functions on $\{0, 1\}$ defined by the truth tables for conjunction, disjunction, and negation. On the lattice-theoretic approach we define a $\leq$ relation on $\{0, 1\}$ as follows: for all $x, y \in \{0, 1\}$, $x \leq y$ iff an arbitrary conditional $p \to q$ is true whenever p has value x and q has value y. So $1 \leq 1$, $0 \leq 1$, and $0 \leq 0$. But $1 \not\leq 0$. Then,

Theorem 1. $(\{0, 1\}, \leq)$ is a boolean lattice.

Empirically the truth value of a conjunction of Ss is the glb of the truth value of its conjuncts, that of a disjunction the lub of those of its disjuncts, and that of not p the complement of that of p.

Lastly, the denotation sets for expressions we have seen are sets of functions whose codomain B is a boolean lattice and whose boolean structure "lifts" to that of the function set:

Theorem 2 (Lifting). If $(B, \leq)$ is a boolean lattice and A any non-empty set, $([A \to B], \leq)$ is a boolean lattice, where $\leq$ is defined by:

$$F \leq G \text{ iff for all } x \in A, F(x) \leq G(x).$$

In writing $F(x) \leq G(x)$, $\leq$ is the partial order relation in B since $F(x)$ and $G(x)$ are elements of B. On the left the $\leq$ is the relation we are defining between functions. This definition is said to be *pointwise* since whether $F \leq G$ depends on their values at the "points" in A. *Provably* also meets, joins, and complements behave pointwise:

Theorem 3. For $([A \to B], \leq)$ as above, $(F \wedge G)(x) = F(x) \wedge G(x)$; $(F \vee G)(x) = F(x) \vee G(x)$ and $(\neg F)(x) = \neg(F(x))$. The zero element in $[A \to B]$ maps every x in A to 0_B the zero of B. The unit maps each x to 1_B, the unit of B.

Theorem 4. If B above is complete so is the lifted lattice: Let j any set such that for each $j \in J$, F_j maps A into B. Then $\{F_j \mid j \in J\}$ has a glb, namely that function mapping each x in A to $\bigwedge\{F_j(x) \mid j \in J\}$. And the lub of that set maps each x to $\bigvee\{F_j(x) | j \in J\}$. These sets are often noted $\bigwedge_{j \in J} F_j$ and $\bigvee_{j \in J} F_j$ respectively.

An example of lifting is the set $[E \to \{0, 1\}]$ in which P_1s denote. To say that **laughed loudly** $\leq$ **laughed** just says that any x that laughed loudly is an x that laughed, which is correct. And empirically, a conjunction of P_1s denotes the glb of the denotations of its conjuncts, the disjunction the lub of its disjuncts, and a negation the boolean complement of the denotation of the expression negated.

Since it is a theorem that $(p \wedge q)(b) = p(b) \wedge q(b)$, etc. we don't have to stipulate that *Ted is laughing and is crying* is logically equivalent to *Ted is laughing and Ted is crying*. It follows from interpreting *and* as the glb operation, as is cross categorically general.

Further, the set in which P_{n+1}s in general denote is recursively a pointwise boolean lattice, the set of maps from E into the set in which P_n's denote. So once we start with $\{0, 1\}$, the set in which P_0s denote, the rest are boolean lattices (BLs) by Lifting.

Recall: earlier we interpreted P_1s in $P(E)$, not in $[E \to \{0, 1\}]$. But we have just seen that these two sets are domains of BLs, the first with respect to $\subseteq$ the second the pointwise $\leq$. And these BLs are isomorphic, noted $\simeq$. By definition $(B, \leq_B) \simeq (D, \leq_D)$ iff there is a one to one function f from B onto D such that for all $x, y \in B$, $x \leq_B y$ iff $f(x) \leq_D f(y)$.

Theorem 5. For any set E, $([E \to \{0, 1\}], \leq) \simeq (P(E), \subseteq)$. The map sending each F to $\{b \in E \mid F(b) = 1\}$ is an isomorphism. Also, the map sending each subset of A of E to the function that maps just the elements of A to 1 is an isomorphism in the other direction.

Theorem 6 (See Enderton 1972, p. 92). Isomorphic structures make the same sentences true (and are said to be *elementarily equivalent*).

Elementary equivalence shows that we can be indifferent as to whether we say $b \in p$ or $p(b) = 1$. Treating a P_1 as a subset of E or a function mapping E into $\{0, 1\}$ will not change which sentences stand in the entailment relation and which expressions denote objects in the appropriate $\leq$ relation. And what holds of P_1s here extends to P_ns: $(P(E^n), \subseteq) \simeq ([E^n \to \{0, 1\}], \leq)$.

Another instance of a lifted algebra is our set of type (1) functions given by $[P(E) \to \{0, 1\}]$. DPs like *every student*, etc. denote in this set, but not uncommonly in the linguistic literature one says that DPs denote sets of properties, i.e. elements of $P(P(E))$. Here we use the functional approach.

Back to quantifiers The DPs we have considered have their values at an n+1-ary relation determined by the functions in $[P(E) \to \{0, 1\}]$, and, as expected, their values are determined pointwise, with *and*, *or*, and *not* denoting the expected boolean operators.

(1)   a.   Most students and almost all teachers are liberal

      b.   Not more than ten students came to the party

      c.   Either a linguist or some deranged engineer designed that

Worth noting is that English has a variety of ways of expressing boolean structure without using *and*, *or*, and *not* directly. (Below, upper case indicates contrastive stress):

(2)   a.   At most ten students ≡ not more than ten students

      b.   Between five and ten cats ≡ at least five cats and not more than ten cats

      c.   Neither every student nor every teacher ≡ not every student and not every teacher

      d.   Only SOME students ≡ some but not all students

And given that DPs denote in a BL we see that expressions of type (1,1) denote in the obvious pointwise lattice: $[P(E) \to [P(E) \to \{0, 1\}]]$.

And we do indeed find boolean compounds of Dets:

(3) a.   some but not all / most but not all

    b.   either hardly any or else almost all

    c.   at least ten but not more than twenty

    d.   either between one and ten or between forty and fifty

And again natural boolean interpretations are available without the use of the standard boolean expressions *and*, ...

(4) a.   (Between five and ten) (students) ≡ (at least five and not more than ten)(students)

    b.   (At most half the)(boys) ≡ not more than half the boys

    c.   Neither less than a quarter nor more than a half...

    d.   Exactly ten/half ≡ at least ten/half and not more than ten/half

Even the quantifiers of type $((1,1),1)$ denote in a pointwise BL: $[P(E) \times P(E) \to [P(E) \to \{0,1\}]]$ and form boolean compounds:

(5) a.   More cats than dogs but not more chickens than turkeys (were inoculated)

    b.   No more poems than plays were on the reading list

The syntactic ubiquity of boolean compounds actually constitutes a major linguistic generalization:

Gen 1. Most categories of English expression form and interpret boolean compounds in *and/but, or, not, neither... nor...*. So called "functional categories" like number and agreement markers are usually not boolean.

    Why is Gen 1 unsurprising? We have an answer: the sets in which most categories of expressions denote are sets with a boolean structure (whether semanticists choose to refer to it or not), so boolean compounds are guaranteed to be interpretable. This also accounts for why *and* and *or* are easily borrowed by languages which lacked them historically. It also accounts for why we can interpret nonce boolean compounds even when they are clumsy or unexpected. We do not think for example of Prepositions (*at, in,* etc.) as particularly boolean, yet *Ted lives neither in nor near New York City* is unproblematic in interpretation.

    In a more speculative vein we might consider that the boolean operations expressed by *and, or, not, neither... nor...* represent properties of mind, or as Boole had it "laws of thought". Certainly as we have seen, their meaning is not specific to any one grammatical category (and there are boolean compounds in many categories we have not yet considered — Adjectives: *an attractive but not very well built house*; Transitive verbs: *John both praised and criticized each student*; Adverbs: *He works slowly but not carefully*, etc. The syntactic ubiquity of boolean compounds suggests that the boolean operations express more how we think about things than they do properties of the things we think about.

Generative grammarians have a tendency to assume that non-sentential uses of boolean connectives can be derived by reduction operations from S-level ones. But as we have seen, the sets in which most expressions denote have a boolean structure, so why not use it? It would seem deceptive on the part of natural language to allow us to form expressions which could not be directly interpreted. One recalls here one of Einstein's favorite adages:

(6)   **Raffiniert ist der Herrgott aber Boshaft ist er nicht**
       Subtle is the Lord but mean he is not

Also, we shall see that several classes of natural language expressions — Predicates & Arguments, Determiners, Modifiers (Adjectives, Adverbs) can be discriminated by the boolean properties of their denotations.

Anticipating an extension of Gen 1 we note that the logical quantifiers **all** and **some** are semantic generalizations of **and** and **or**. *And* combines two, even n-many conjuncts ($n \geq 2$) and denotes the glb of their denotations. *Or* is analogous, denoting their lub. And, as we see shortly, **all** and **some** are just arbitrary glb and lub operators, not limited to the finite case (but including them). E.g. in a model in which Ed, Ted, and Ned are the only students, *Ed and Ted and Ned* has the same denotation as *every student*. And *Ed or Ted or Ned* denotes the same function as *some student*. Further we'll see that the extension of Gen 1 covers much more than just **all** and **some**. But first we treat proper nouns and the "individuals" they denote.

Proper Nouns and Individuals   Of the denotation sets we have considered the universe E stands out as the only one lacking a designated boolean relation. It is just an arbitrary set. But if we treat proper nouns — *John, Rosa*, etc. as individual constants (of type e) as in FOL they will denote in E.

Yet proper nouns form boolean compounds with each other — *Rosa and Zelda are ballet dancers, Neither Rosa nor Zelda laughs at my jokes*. This is unnatural if there is no boolean relation on E. Proper Nouns also form boolean compounds with quantified DPs as in (7). Again this is unexpected if proper nouns and quantified DPs denote significantly different kinds of objects.

(7) a. John and some student (were taking tickets when I arrived)

  b. Neither Frank nor any teacher (arrived early)

  c. Ah, so Sue and not Ann (will represent us at the meeting)

Note that in isolation *Not Ann will represent us...* is ungrammatical. So *Sue will represent us at the meeting and not Ann will represent us at the meeting* is implausible as a derivational source for (7c).

So we would like to find some way that *John* and *some student*, etc. can denote in the same set. We cannot force quantified DPs to denote in E since, for E finite, the number of logically distinct DPs, which include quantified ones and their boolean compounds, vastly outnumber the elements of E. If $|E| = n$ (n may be infinite here, but convincing cases arise for small finite n) then $|[E \to \{0,1\}]| = |P(E)| = 2^n$, as in general $|[A \to B]| = |B|^{|A|}$, and so $|[P(E) \to \{0,1\}]| = 2$ raised to the power 2^n. So in a model with just 4 objects there are $2^4 = 16$ possible P_1 denotations, and $2^{16} = 65,536$ possible DP denotations!

Now, there is a natural way to "type raise" entities in E to functions from $P(E)$ into $\{0,1\}$, due ultimately to Montague (1973) and Lewis (1970): in Chapter 12 we propose a sleeker alternative, which maintains the logical effect of type raising and generalizes more naturally to some new intensional phenomena.

Definition 4. For each $b \in E$, I_b is$_{\text{def}}$ that function from $P(E)$ into $\{0,1\}$ given by: $I_b(p) = 1$ iff $\{b\} \subseteq p$, all $p \subseteq E$. (NB: $\{b\} \subseteq p$ iff $b \in p$.)

I_b is called an *individual*, the one *generated* by b. If *John* denotes an individual I_j, $I_j(\textbf{doctor}) = 1$ iff $j \in \textbf{doctor}$, the same truth conditions as obtained when *John* is interpreted as j in E. Now we can interpret the boolean compounds of proper nouns and quantified DPs in (8):

(8) a. John and some student $= I_j \wedge \bigvee \{I_b \mid b \in \textbf{student}\}$

  b. Neither Frank nor any teacher $= \neg(I_f \vee \bigvee \{I_b \mid b \in \textbf{teacher}\})$

  c. Sue and not Ann $= I_s \wedge \neg I_a$

(As in *So Sue and not Ann will represent us at the meeting*)

The boolean representations of complex DPs in terms of individuals in (8) generalizes in a surprisingly strong way:

Theorem 7 (Weak Generator Theorem). Every function H from $P(E)$ into $\{0,1\}$ is a boolean function of individuals.

This just means that every such H is expressible as some combination of meets, joins and complements of individuals, as illustrated below.

So IND_E, the set of individuals over E, *generates* $[P(E) \to \{0,1\}]$. Here is a proof sketch:

Let H any function from $P(E)$ into $\{0,1\}$ and for each $p \in P(E)$, write f_p for that function from $P(E)$ into $\{0,1\}$ which maps p to 1 and all other properties q to 0. Then $H = \bigvee \{f_p \mid H(p) = 1\}$. So we form the set of f_p, for p a property that H holds of, and we take the lub of that set. The equation holds since $\bigvee \{f_p \mid H(p) = 1\}$ holds of some q iff f_q is in the set we are taking the lub of, and that is so iff $H(q) = 1$.

So we see that H is a boolean function, lub ($\bigvee$), of functions of the form f_p. We show that any f_p is a boolean function of individuals. Now f_p holds of a q iff q is p. And *all(p) and no non-p* holds of q iff q is p. And this latter is: $\bigwedge \{I_b \mid b \in p\} \wedge \neg \bigvee \{I_b \mid b \in \neg p\}$, i.e. all individuals with p and not [some individuals with non-p]. Thus $H = \bigvee \{\bigwedge \{I_b \mid b \in p\} \wedge \neg \bigvee \{I_b \mid b \in \neg p\} \mid H(p) = 1\}$ and so is a boolean function of individuals.

The Weak Generator Theorem can be enlighteningly strengthened in a way of some possible linguistic interest. Namely IND_E is a set of independent (i.e. *free*) generators for $[P(E) \to \{0,1\}]$, the set of generalized quantifiers. Loosely this means that the behavior of one individual does not influence that of another. Formally:

Theorem 8 (Free Generators). Any map from IND_E into any complete atomic (ca) BL extends to a complete homomorphism (The Justification Theorem of Keenan and Faltz 1985).

A homomorphism h is *complete* iff it commutes with arbitrary glbs and lubs: $h(\bigwedge_{j \in J} F_j) = \bigwedge_{j \in J} h(F_j)$, etc. The theorem says that we can transfer individuals *in any way we like* to any other ca BL and their images behave booleanly as they did as generalized quantifiers.

Observe that the map sending each b in E to I_b is one to one onto the set of individuals, so the number of individuals, $|\mathrm{IND}_E| = |E|$. But the total number of possible DP denotations is 2 raised to the power $2^{|E|}$. So in a model with just 4 individuals, all 65,536 functions from properties to truth values are expressible as boolean functions of individuals. Cool!

In passing we note that early (and much not so early) work in linguistics treated Proper Nouns as the dominant arguments of verbs. In a sense Theorem 8 accounts for that, given that boolean operations are cognitively inherent. Then given Proper Noun denotations, individuals, we cognitively capture all subject DP denotations as boolean functions of the functions denotable by Proper Nouns.

And we can now support our earlier remark that **all** and **some** are generalized glb and lub operators. Consider $\bigwedge\{I_b \,|\, b \in \mathbf{doctor}\}$. By pointwise meets it holds of a property q iff each individual in the set holds of q, that is, each individual with the doctor property. And this just says that the b's which generate those individuals are in q, so they form a subset of q. That is, $(\bigwedge\{I_b \,|\, b \in \mathbf{doctor}\})(q) = \mathbf{all}(\mathbf{doctor})(q)$. And in general then, $\mathbf{all}(p) = \bigwedge\{I_b \,|\, b \in p\}$.

Similarly $\mathbf{some}(p) = \bigvee\{I_b \,|\, b \in p\}$. That is, $\mathbf{some}(p)$ holds of q iff at least one individual with p has q, that is, $p \cap q \neq \varnothing$. Thus we see that quantifiers in FOL generalize the $\wedge$ and $\vee$ operations of SL.

A Speculation and a Result on Logical Expressive Power   The speculation first: We have noted that early work in generative grammar derived pairs of Ss that differed by the presence of a DP. E.g. *Ed is laughing and (is) crying* vs *Ed is laughing and Ed is crying*. The DPs used were usually individual referring ones, like proper nouns, etc. It was assumed the meaning of the Ss was determined on the underlying "deep" structure from which they were derived, so the Ss were predicted to be paraphrastic.

But, as noted earlier, once quantified DPs were considered the Ss were typically not paraphrases: *Some child is laughing and (is) crying* $\not\equiv$ *Some child is laughing and some child is crying*.

Query: Why did linguists take proper nouns as representative of DPs (then called NPs for *Noun Phrases*) in general? The speculation: it was because proper nouns (or singular definite DPs, like *the man*) denoted individuals, and modulo the boolean operations all denotations of (count) DPs are in fact boolean compounds of individuals. Proper Nous are not *syntactic* generators of DPs of course — DPs like *some/all/most poets* do not even contain a proper noun. But individuals are semantic generators, so if you apprehend what an individual is and your boolean competence is quite general, not limited to DPs certainly, then in some sense you apprehend the possible denotations of all count DPs.

An alternative explanation that seems "natural" but is empirically incorrect is that proper nouns are the most frequent DPs in discourse and texts. But a recent study by Wymark (2018) shows that only 14.1% of DPs in the Wall Street Journal corpus are headed by singular proper nouns. The total number of DPs counted was 432,777. Another 5% were personal pronouns (many of which were surely anaphoric rather than individual referring). But 25.8% of DPs were of the form Det+(X)+N+(Y), where Det includes possessives. In addition 34.2% were DPs headed by bare common nouns such as *securities firms*, painted in *oils*, *On the issue of abortion*. So, while refinement of the categories is on-going it seems clear that DPs which are quantified, including bare DPs, outnumber Proper Nouns by about four to one.

An expressivity result that supports our psycholinguistic claim is: Given E finite, every H from $P(E)$ into $\{0,1\}$ can be denoted. We see shortly that for Adjectival Modifiers and Dets (Type (1,1)) we do not have complete expressivity. But for DPs we do. For suppose E is infinite and that $I_0, I_1, I_2, \ldots$ is a denumerable sequence of individuals. Let *Dana* denote I_0 and interpret the P_2 *admire* as $\{\langle I_{n+1}, I_n \rangle \mid n > 0\}$. Then *Dana, the individual who admires Dana, the individual who admires the individual who admires Dana*, etc. denote the members of this sequence, so each finite subset is denotable by a finite disjunction. Then each f_p is denotable, as it is **all**(p), a finite meet of denotable individuals, meet the complement of the join of a finite set of denotable individuals. Thus, since H itself is a finite join of f_p, it is denotable.

As we consider Dets and modifying (not predicative) Adjectives (functions from $P(E)$ to $P(E)$) we find that there are, possibly related, non-trivial constraints on what we can denote.

A Semantic Universal of Adjectives First some examples of modifying adjectives:

(9) a.   a female lawyer, a bald soprano, a 6 foot tall student

   b.   a tall student, a large jockey, an expensive house

   c.   a skillful pianist, an able flautist, a famous tennis player

   d.   an apparent millionaire, an alleged murderer

   e.   a potential candidate, a former president

The range of semantic dimensions along which adjectives can vary is enormous (Burnett 2014; Rotstein and Winter 2004; Vendler 1968), the sample above is in nowise exhaustive. Some involve notions not represented in our naive model theory at all: a *former* president is not a president, but used to be, a *potential* candidate isn't one but might become one.

Still, here we treat adjectives as functions from properties to properties. For the moment think of properties as elements of a complete BL P, not necessarily the set of entities that have those properties. Most typically if F

is an adjectival function and p a property, $F(p) \leq p$. Call F meeting this condition *restricting*. So an $F(p)$ is a subtype of p: a female lawyer is a lawyer, a tall student a student, a skillful pianist a pianist. A few exceptions are in (9d,e). To take Montague's example, an alleged murderer need not be a murderer, it is simply someone who is alleged to be one.

Non-restricting modifiers are always, to our knowledge, *non-extensional*. E.g. if it happens that the murderers and the thieves are the same individuals it, obviously, does not follow that the alleged murderers and the alleged thieves are the same. This is in fact our proposed universal:

(10) **?A Universal**: All extensional modifiers are restricting.

Note that many restricting modifiers (maps from X to X), like those in (9c), fail to be extensional. It could happen that the surgeons and the poker players were the same individuals in some situation, but it wouldn't follow that the skillful surgeons and the skillful poker players were the same. (See Chapter 12 for a semantic analysis.)

The restricting adjectives in (9a,b), while logically somewhat different are restricting and extensional. We define:

Definition 5. For P a (complete) BL, $(\{F \in [P \to P] \mid F(p) \leq p\}, \leq)$ is a (complete) BL where $\wedge$ and $\vee$ are pointwise and $(\neg F)(p) = p \wedge \neg(F(p))$. The zero is the pointwise 0 and the unit is the identity map: $P \to P$.

So for example a tall and handsome doctor is someone who is a tall doctor and also a handsome doctor; a not very tall doctor is a $\underline{doctor}$ who fails to be a very tall one (not simply any object, like my cat, which also fails to be a very tall doctor). Such algebras will be noted $\mathrm{Rest}(P)$ for "restricting functions from P to P". They are particular cases of *factor algebras*, the set of elements in a BL which are $\leq$ some fixed $x > 0$. In this case restricting $F \leq \mathrm{id}_P$, the identity function from P to P.

For $P = P(E)$ with $|E| = n$ that there are $\prod_{1 \leq k \leq n} C(n,k) \cdot 2^k$ elements in such algebras. So being restricting is a strong condition on the functions allowed, as the total number of functions from $P(E)$ to $P(E)$ is 2^n to the power 2^n which is (2 to the $(n \cdot 2^n)$). So if $|E| = 3$ there are (2 to $(3 \cdot 2^3)$)) $= 2^{24} > 16$ million maps from $P(E)$ to $P(E)$, only 576 of which are restricting. $[P(E) \to P(E)]$ is hyperexponential in n but $|\mathrm{Rest}(P(E))|$ is merely exponential in n.

Now the adjectives in (9a) are *absolute*, a special case of being restricting:

Definition 6. F from $P(E)$ to $P(E)$ is *absolute* iff $F(p) = p \cap F(E)$.

E.g. a female lawyer is a lawyer who is a female individual; a six foot tall student is a student who is a six foot tall individual. In general relative clauses formed on extensional positions are absolute: a doctor I shook hands with is a doctor who is an individual I shook hands with. And absolute F are restricting, since $p \cap F(E) \subseteq p$.

Theorem 9. Abs(E), ($\{F \in [P(E) \to P(E)] \mid F$ is absolute$\}, \leq$), is a complete sublattice of Rest($P(E)$).

$(D, \leq)$ is a *complete sublattice* of $(B, \leq)$ iff $D \subseteq B$, the $\leq$ relation on D is that of $(B, \leq)$ restricted to D, and for $K \subseteq D$, the glb of K in $(D, \leq)$ is the same object as the glb of K in $(B, \leq)$.

Theorem 10. (ABS(E), $\leq$) $\simeq$ ($P(E), \subseteq$). The map h sending each F to $F(E)$ is an isomorphism.

This theorem says that the absolute adjective denotations are, in effect, just the first order one place predicates. So $|\text{ABS}(E)| = 2^{|E|}$. Unsurprisingly then we support the following empirical generalization:

(11)   Absolute adjectives are extensional.

So in a model in which the doctors happen to be the same individuals as the lawyers we can infer that the female doctors and the female lawyers are the same. But extensionality in general fails for mere restricting non-absolute adjectives. In the previous model, the skillful doctors and the skillful lawyers may be different, as may the good doctors and the good lawyers.

And now the *relative* adjectives in (9b), like *tall*, take on added interest. They are restricting: a tall jockey is a jockey. But a tall jockey need not be a tall individual, but only tall compared to jockeys. Notice that *tall* $\not\equiv$ *tall for*. We can say that Bill is awfully tall for a jockey as an argument against his being a jockey, so it certainly does not imply that he is a jockey. Also *tall for* is not extensional: if the jockeys and the basketball players are (unexpected but not a logical impossibility) the same individuals, Bill could be tall for a jockey but not so for a basketball player. *Tall for an N* implies a standard or expectation regarding the height of Ns, regardless of whether there are any or not. So whether Bill is tall for an N depends on more than just the height of the Ns, but also on our standards or expectations.

Returning now to *tall*, etc. in (9b) they do seem to be extensional: if the doctors and the lawyers are the same then so are the tall doctors and the tall lawyers as we are comparing heights within the same set.

So the restricting functions they denote take us well outside of first order. The number of restricting functions that are not absolute is $2^{(n+1)(n+2)/2} - 2^n$ which, for $|E| = n = 4$, computes to $32,768 - 16 = 32,752$. So almost all restricting functions are not absolute, and:

(12)   Adjectives like *tall* take us outside first order expressive power.

We note too, though beyond the scope of this work, that relative adjectives form comparatives naturally and that comparatives of extensional adjectives seem, on cursory examination, to be extensional and those built from properly intensional ones intensional: Assume a situation in which the doctors and the lawyers are the same individuals. Then *Ed is a taller doctor than Al* iff *Ed is a taller lawyer than Al* but *Ed is a smarter doctor than Al* and *Ed is a smarter lawyer than Al* may have different truth values. This is, naively, true because *x is a taller p than y* iff *x* and *y* are both in *p* and *x* is a taller individual than *y*. This fails if *tall* is replaced by *skillful*. So we support:

(13)   The comparative **F-er than** / **as F as** preserves extensionality:
       F is extensional iff **F-er than** and **as F as** are

A Semantic Universal concerning Determiners   Determiners are somewhat like "higher order" adjectives (Indeed Dets like *this*, *those*, etc. are traditionally called *demonstrative adjectives*): they both take properties as arguments, but adjectives yield as value a single property, while Dets yield a set of properties (equivalently a function from properties to truth values) as values. And limiting ourselves to extensional Dets (as we have for adjectives so far) we see that the (first) property argument of both bounds the value in a similar way. For extensional adjectives, $F(p)$ is always a subset of p. For Dets the value of F at p is determined by a set of subsets q of p.

Definition 7 (*Conservativity*). A function $F \in [P(E) \to [P(E) \to \{0,1\}]]$ is *conservative* iff for all $p, q \in P(E)$, $F(p)(q) = F(p)(p \cap q)$.

(This is the usual definition: given as an invariance condition as we did for the (co-)intersective and proportional Dets it would read: for all p, q, q' if $p \cap q = p \cap q'$ then $Fpq = Fpq'$.) We write Cons_E for the set of (one place) conservative functions over E.

The Conservativity Universal (Cons)
Extensional Dets denote conservative functions

Cons yields the judgments of logical equivalence below:

(14) a. Some/All/Most students sleep late ≡ Some/All/Most students are students and sleep late

   b. More male than female children shout a lot
       More male than female children are children and shout a lot

   c. Most of Ted's articles were accepted
       Most of Ted's articles were articles that were accepted

The judgments of equivalence are banal here as the predicate in the second sentence just repeats information already given in the noun.

Conservativity is properly stated for n-place Dets in the "obvious" way: an n-place Det F is conservative iff its value at n-tuples p and q map each property r to the same truth value provided $p_i \cap r = q_i \cap r$, all $1 \leq i \leq n$. See Keenan and Moss (1985) for more discussion. Here we just focus on Det$_1$s for simplicity of presentation. And we note that meets, joins, and complements of cons functions are cons, and Cons is a very strong constraint on possible Det denotations (Theorem 12):

Theorem 11. (Cons$_E$, $\leq$) is a complete sublattice of ([$P(E) \to \{0,1\}$], $\leq$).

Theorem 12. Of the 2 to the 4^n maps from $P(E)$ into [$P(E) \to \{0,1\}$], only 2 to 3^n are conservative (Keenan and Stavi 1986).

An example of a non-conservative function is D defined by:

(15) $D(p)(q) = 1$ iff $|p| = |q|$ (or   ...iff $|p| > |q|$, ...)

Cons fails here since e.g. it may happen that $|p| = |q|$ but $|p| \neq |p \cap q|$. We can of course express equi-cardinality in English: *Exactly as many ps as qs exist*, where *exist* denotes E (or the property that maps all $x \in E$ to 1), but we cannot express it with a one place Det. See Beghelli (1994) for figures for the Cons functions in higher types.

To see that Cons is very strong we note that in a universe of just two individuals there are $2^{16} = 65,536$ functions from properties to generalized quantifiers, only 512 of which are conservative (!). And of those 512 only 30 lie in INT ∪ CO-INT (and smaller sets of generators can be found).

Keenan and Stavi (1986) also support a strong result in the other direction. Namely, for finite E every conservative function of type (1,1) is denotable in the sense that given a cons F we can form a complex Det$_1$ expression which can be interpreted as F. This depends on the (modest) freedom a model has in assigning denotations to proper nouns, possessives or extensional adjectives. The Det_1 expressions themselves are cumbersome and ungainly. But they make the point that we cannot strengthen Cons, as then we would rule out some interpretations that we want for certain Dets.

What motivates Conservativity? Probably one's most immediate response is cognitive: the noun argument of a Det delimits the class of things we are talking about. (But see below.) In classical mathematical languages — Elementary arithmetic, Euclidean geometry — there is an intended class of objects we use the language to talk about. But more typically mathematical languages — that of group theory, point set topology, boolean algebra, etc. are designed to let us speak of any collection of objects on which are defined relations (functions) satisfying certain axioms. These axioms typically have many non-isomorphic models. But daily discourse countenances no axiom sets which we speak to satisfy.

Natural languages are general purpose, used to speak about anything at all. So we need ways to specify what we are talking about and quantifying over "on line" as it were, at the moment of utterance. Using common noun phrases satisfies this need.

Keenan and Stavi (1986) offered a different, not incompatible, motivation for conservativity. We wanted certain functions to be in our denotation set for Dets because we had fairly simple expressions (*some*, *several*, *all*, *his*/*John's*, *at least five* (and later *most* and *half*) which denoted them. And we wanted to be able to form and interpret boolean compounds like *some but not all*, *at least two and not more than ten*, etc. When we formulated more precisely the initial functions we wanted and then added in all those we could get by iterated application of the boolean operations ($\wedge, \vee, \neg$) we provably obtained exactly the conservative functions. Here is a more up to date statement (definition to follow):

Theorem 13. The complete boolean closure of $\mathrm{INT}_E \cup \mathrm{CO\text{-}INT}_E = \mathrm{CONS}_E$. (Keenan 1993, p. 317)

This expresses the intuition that if you start with just the intersective and co-intersective functions (Each equi-numerous with the generalized quantifiers) and then add in all those obtainable by iterative formation of boolean functions ($\wedge, \vee, \neg$) of those you already have, the result is exactly the set of conservative functions. The proof of Theorem 13 requires some work, which we give here.

Lemma 1. INT_E and $\mathrm{CO\text{-}INT}_E$ are complete sublattices of CONS_E.

We note as well that CARD_E is a complete sublattice of INT_E. When $|E| = n$ is finite the atoms (Def 9 below) of CARD_E are the functions $\mathbf{ex}(k)$, for $0 \leq k \leq n$, where $\mathbf{ex}(k)(p)(q) = 1$ iff $|p \cap q| = k$. So $|\mathrm{CARD}| = 2^{n+1}$. Analogous claims hold for $\mathrm{CO\text{-}CARD}_E$.

Lemma 2. If $\{Y_j \mid j \in J\}$ is a family of (domains of) complete sublattices of a boolean lattice B then so is $\bigcap\{Y_j \mid j \in J\}$.

[If $K \subseteq \bigcap\{Y_j \mid j \in J\}$ then for each j, $K \subseteq Y_j$, so $\bigwedge K \in Y_j$ since Y_j is complete, whence $\bigwedge K \in \bigcap\{Y_j \mid j \in J\}$, so it is also complete.]

Definition 8. If $K \subseteq B$, a boolean lattice, then $\mathrm{BC}(K)$, the *complete boolean closure* of K, $= \bigcap\{K' \subseteq B \mid K \subseteq K' \ \& \ K'$ is (the domain of) a complete boolean sublattice of $B\}$.

Theorem 14. $\mathrm{BC}(K)$ as above is a complete sublattice of B. It is the least one that includes K, i.e. it is a subset of any complete sublattice of B which includes the elements of K.

The proof of closure above uses the important notion of an *atom*:

Definition 9. For $(B, \leq)$ a boolean lattice and $\alpha \in B$,

    a. α is an *atom* iff for all $x \in B$, $x \leq \alpha \Rightarrow (x = 0 \,\mathrm{or}\, x = \alpha)$.
    b. $(B, \leq)$ is *atomic* iff for all $\beta \neq 0$, there is an atom α with $\alpha \leq \beta$.

So an atom is a smallest non-zero element. Only the zero is strictly less than it (where $x < y$ iff $x \leq y$ and $x \neq y$). E.g. in a power set algebra the atoms are the one member sets, the zero element being the empty set. Also power set algebras are atomic, since if $B \neq \varnothing$ there is a $b \in B$ and the atom $\{b\} \subseteq B$. So all power set algebras are complete and atomic. In fact we have a converse:

Theorem 15. P is complete and atomic (ca) $\Rightarrow$ P is isomorphic to $(P(\mathrm{Atom}(P), \subseteq)$, the power set of the set of its atoms. The map sending each x to $\{\alpha \in \mathrm{Atom}(P) \mid \alpha \leq x\}$ is an isomorphism.

Corollary. P is ca iff P is isomorphic to a power set BL.

Worth noting is that all finite BLs are complete and atomic (taking as axiomatic that $0 \neq 1$) and are thus power set lattices up to isomorphism. Of utility in what follows are the next two theorems:

Theorem 16. B is atomic $\Rightarrow$ for all $x \in B$, $x = \bigvee\{\beta \in \mathrm{Atom}(B) \mid \beta \leq x\}$.

In this theorem B need not itself be complete. But the set of atoms that lie below a given element x always has a lub, namely x itself.

Corollary. $[P(E) \to \{0,1\}]$ is atomic. (The functions f_p, $p \in P(E)$, which map a q to 1 iff $q = p$ are the atoms. These are the f_p used in the Generator Theorem earlier.)

Theorem 17. If $(B, \leq_B)$ is complete and atomic and $(A, \leq)$ is a complete sublattice of B then A is also atomic. $\mathrm{Atom}(A) = \{\bigwedge\{x \in A \mid \beta \leq_B x\} \mid \beta \in \mathrm{Atom}(B)\}$.

Theorem 18. INT_E, $\mathrm{CO\text{-}INT}_E$ and CONS_E are all atomic.

 a. The atoms of INT are given by: $f_p(s,t) = 1$ iff $s \cap t = p$, all $p \in P(E)$
 b. The atoms of $\mathrm{CO\text{-}INT}$ are: $g_p(s,t) = 1$ iff $s - t = p$, all $p \in P(E)$, and
 c. The atoms of CONS are the $f_{p,q}$, for $q \subseteq p$, where $h_{p,q}(s)(t) = 1$ iff $s = p$ and $s \cap t = q$

Clearly f_p is intersective and since it maps one property to 1 it is not 0_{INT}, and since it maps just one property to 1 only 0_{INT} is strictly less than it, so it is an atom. Similarly g_p is an atom in the co-intersective lattice . And $h_{p,q}$ is an atom in CONS as it maps only p to a non-zero generalized quantifier, one that can't distinguish between properties t and t' that have q as their intersection with p, so it is conservative.

Now to see why Theorem 13 holds observe that for $h_{p,q}$ an atom of CONS, $h_{p,q} = f_q \wedge g_{p-q}$, the meet of an intersective atom and a co-intersective atom, hence in the closure of $\mathrm{INT} \cup \mathrm{CO\text{-}INT}$. And since the closure is complete and every conservative function is the join of its atoms (Theorem 16) all conservative functions are in the closure. And all functions in the closure are conservative since the F in $\mathrm{INT} \cup \mathrm{CO\text{-}INT}$ are, and conservativity is preserved by $\wedge$, $\vee$ and $\neg$. Thus identity holds.

Other constraints on Det interpretations have been tried but mostly found wanting (van Benthem 1986, Ch 1). Chapter 10 discusses a global generalization of Det interpretations.

Non-conservative Determiners? Several candidates for English Dets that are not conservative have been proposed. *Only* was quickly noticed by many. The first in print as I recall is Johnsen (1987) who also noticed *mostly*. The (a,b) pairs below are logically equivalent:

(16) a. Only poets daydream a'. All daydreamers are poets
   b. Mostly poets daydream b'. Most daydreamers are poets

Thus **only** behaves as the converse of **all** and **mostly** as that of **most**. We note that (16a) may be true or false depending on how the world is, but *Only poets are poets who daydream* is trivially true, so not equivalent to (16a), thus *only* is not conservative. A similar claim holds for (16b). Treating *only* and *mostly* here as type (1,1) functions their semantics is given by:

(17)   a.   **only**$(p)(q) = 1$ iff **all**$(q)(p)$, i.e. $q \subseteq p$

      b.   **mostly**$(p)(q) = 1$ iff **most**$(q)(p)$, i.e. $|q \cap p|/|q| > \frac{1}{2}$

Of these two *only* has been better documented. Certainly it has many non-Det uses, combining with full DPs (*Only Ann laughed*), VPs (*He only sang, he didn't dance*), and seemingly other Dets *Only a few showed up*. It shows up as a postposition on California freeways: *Exit Only*. And consider the distributional and semantic differences in the use of *only* in (18):

(18)   a.   Only the two students who aced the exam won scholarships.

      b.   The only two students who aced the exam won scholarships.

Cross linguistically *only* is also ubiquitous. All 36 of the languages covered in KP/PK had a word or phrase naturally translated as *only*.

    Clearly these adverbial looking Dets are in need of more syntactic and semantic study. But one property they have with more commonly cited Dets is that they can be repeated in argument position of the same predicate, which discourages a view on which they are sentence level operators:

(19)   a.   All poets like all poets.

      b.   Only poets like only poets.

      c.   Mostly poets like mostly poets.

More recently Zuber (2018) notes these and adds some new candidates, suggesting that this pattern is not simply idiosyncratic to *only* and *mostly*. Consider:

(20)   a.   Also (some) teachers danced.

      b.   **Also**$(p)(q) = 1$ iff $p \cap q \neq \varnothing$ and $(\neg p) \cap q \neq \varnothing$

Also+CNP repeats a little less well than *only* and *mostly*, but (21b) seems natural enough (best with slight focus on *John*):

(21)   a.   ? Also John enjoys also baseball.

      b.   Also John also enjoys baseball
             $\equiv$ John enjoys baseball in addition to something else and someone besides John does too.

(Remarkable how much propositional content can be packed into a single word.)

Zuber notes a last intriguing but harder to accept non-conservative Det, per (22b,c). I have italicized the expression intended as a Det. The Polish example (23) is more natural (Zuber 2004).

(22)   a.   Every dancer except Leo is a student.

      b.   *Apart from Leo only* students are dancing.

      c.   *Apart from two teachers only* students are dancing.

(23)   Oprócz Kazia sami   studenci tańcza
      Apart   Kazia "only" students are-dancing
      Except for Kazia only students are dancing

(24) $(\textbf{apart-only}_{p,n})(s)(t) = 1$ iff $|p \cap t| = n$ & $|t| \geq 2$ & $t \subseteq (p \cup s)$

The "Det" in (24) is non-conservative by the last clause, essentially the same problem as with **only**.

Zuber also reminds us of Westerståhl's (1985) reverse proportional reading of *Many Scandinavians have won the Nobel Prize*: the proportion of Scandinavian prize winners is high relative to the number of awards (not the number of Scandinavians).

Finally Zuber finds a generalization that captures all of these cases: they are not classically conservative, that is, conservative on their right (or second) argument but rather conservative on their left (or first) argument. Formally,

Definition 10. *F* of type (1,1) is

    a.   *Right-conservative* (cons2) iff $F(p)(q) = F(p)(p \cap q)$ and
    b.   *Left-conservative* (cons1) iff $F(p)(q) = F(p \cap q)(p)$

By symmetry $|\text{CONS2}| = |\text{CONS1}|$, the function mapping each cons2 function to its converse is an isomorphism. And it is easy to see that the functions that are both cons1 and cons2 are exactly the intersective ones.

Theorem 19. $\mathrm{CONS1} \cap \mathrm{CONS2} = \mathrm{INT}$

And Zuber proposes as a semantic universal that all natural language Dets are *weakly conservative*, meaning either cons1 or cons2. Moreover this is a very strong constraint. Recall that the total number of functions of type (1,1) is 2 to the power 4^n. The classical conservative ones, i.e. cons2, number just 2 to the 3^n and INT is just 2 to the 2^n. So the total number of weakly conservative functions is 2 to the $3^n + 1$ less 2 to the 2^n, as the intersective functions were counted twice.

So in a model with just two individuals there are $2^{16} = 65,536$ functions of type (1,1), just $1024 - 16 = 1008$ of which are weakly conservative. So weak conservativity allows **only**, etc. but still rules out almost all the type (1,1) functions!

Adverbial quantification   I close this chapter by acknowledging the existence of a rich array of predicate level quantifiers, illustrated in (24) with frequency adverbials:

(25)   never, rarely, seldom, twice, sometimes, usually, frequently, often, (almost) always

I have not attempted a formal semantics for such quantificational expressions, but a first study which shows that several basic concepts from the nominal quantifiers we have studied carry over to predicate level ones, which involve time, event, motion, ... is de Swart (1996).

Chapter 8

Characterizing Syntactic Categories Semantically?

Can we characterize syntactic categories of English expressions in terms of the boolean structure of the sets in which they denote? Surely not, at least not completely. There is much in natural language semantics that is not specifically boolean. But we can I think make some inroads to see how some categories are distinguished in boolean terms. This may help us get a grip on the possible semantic architecture of grammatical categories in natural language. The relatively few grammatical categories we have investigated do present diverse boolean structures. And possibly the many categories we do not study here can be at least partially understood as functions or refinements of those we can say something substantive about.

So far we have really just discussed Predicates and their Arguments, Common Noun Phrases (CNPs), Determiners (Quantifiers), and Modifiers.

Predicates So far we have treated n-place predicates, P_ns, in the shallowest of terms, namely simply by the number of arguments they take. Linguists of course want significant refinement here. As we have noted, *to respect* and *to criticize* differ significantly in that one, naively, determines a state and the other an activity. And beyond mere activities we want to distinguish accomplishments — *draw a circle in the sand*, from achievements — *reach the other side*. And within these subcategories we want to distinguish the semantic roles that the different arguments bear: Agent, Patient, Recipient, etc. See Jackendoff (1990) for illustrative discussion.

That said, P_ns, including P_0s (Sentences) do possess a distinguished boolean structure — their denotation sets are complete atomic boolean lattices (BLs). These two properties vary independently: there are BLs which are neither complete nor atomic, ones that are complete but not atomic, ones that are atomic but not complete, and finally, ones that are both complete and atomic (ca).

We have noted that all finite BLs are ca, so being ca is in some sense the unmarked case. In Chapter 12 we use one non-atomic lattice to represent a class of non-extensional modifiers.

Arguments   We have treated count nominal phrases like *most poets* as functions taking $n > 0$-place predicates as arguments to yield predicates with one fewer argument. Linguists would nonetheless refer to *most poets* in *Most poets daydream* as the subject argument of *daydream* and here we will follow that usage when convenient.

We treated arguments in the linguists' sense as denoting generalized quantifiers (GQs), functions from P_{n+1} denotations to P_n denotations, those whose values were determined by their values at P_1 denotations, that is, functions from $P(E)$ into $\{0, 1\}$. We shall often just write P for $P(E)$, as the latter is up to isomorphism just a complete atomic BL. By Lifting the set of GQs is also a complete atomic BL. The atoms are those functions which map a single property (element of P) to 1 and all others to 0.

The GQ lattice however has more structure than merely being complete and atomic: it is *freely generated* by the set of individuals. The individuals recall are the functions I_b, for $b \in E$, which map a property p to 1 iff $b \in p$ (equivalently, iff $\{b\} \subseteq p$). Algebraically we showed that these functions are precisely the complete homomorphisms from P into $\{0, 1\}$. And we showed that any generalized quantifier is a boolean function of individuals. That is, they generate all the GQs in the sense that the functions we can build up by taking arbitrary pointwise meets, joins and complements of individuals are all the functions from $P(E)$ into $\{0, 1\}$.

And to say that the set of individuals is a set of *ca-free* generators of the GQs is to say that they behave independently one from the other. In formal terms appropriate here:

Theorem 1. Let B any ca boolean lattice and h any function from the individuals in GQ into B. Then there is exactly one complete homomorphism from GQ into B that takes the same values at the individuals as h does.

This is an appropriately standard way of saying that no individual has any influence on how another individual will behave — we can freely map them into an appropriate ca boolean lattice. We note that the P_n lattices are not freely generated — they have no "individuals". So argument expressions differ in boolean character from predicates.

We may wonder whether argument expressions other than those of count nominals have individuals, that is, are freely generated. We see below that various types of nominalized sentence combine with boolean operators, so their denotation set should have a boolean structure. But of just what sort we are unsure:

(1)   a.   Ed believes that I broke the vase but not that I did so on purpose.

b.   He believes either that there is life on Mars or that there isn't.

c.   He believes that either there is life on Mars or there isn't.

d.   John believes whatever Bill says.

(2)   John's leaving early but not Fred's leaving early surprised me.

(3)   He knows who you confide in but not who you date.

Common noun phrases   CNPs serve to provide domains of quantification, which seems to be a logical role different from that of one place predicates. We have nonetheless treated them so far as identical to P_1s, the map sending *poet* to *is a poet* is an isomorphism (ignoring tense and aspect). Surely then some enrichment of our semantics for predicates is indicated.

Modifiers   Modifiers, such as adjective phrases (*a tall and awkward fencer*), relative clauses (*a student who you met at the party*) and Prepositional phrases (*a child in the garden*) are commonly restricting: $F(p) \leq p$. As we have seen absolute modifiers, functions F that map each p to $(p \cap F(E))$ are isomorphic to P_1s. But there seem to be some extensional restricting modifiers that are not absolute, e.g. *tall*. And these induce a richer lattice than the merely absolute ones. The full set of restricting functions properly includes the absolute ones. But of course many functions from $P(E)$ to $P(E)$ are not restricting, e.g. the function which maps each p to E. The weakest restricting function is the identity function: $F_=(p) = p$. So let us define in general:

Definition 1. For B a boolean lattice with $0 < x \in B$, $\downarrow x = \{y \in B \,|\, y \leq x\}$ is a boolean lattice with x as the unit, the $\leq$ relation is just that of B restricted to $\downarrow x$, meet, join, and 0 are the same as in B, and complements are taken relative to x. That is, $\neg_{\downarrow x} y = x \wedge \neg_B y$. This algebra is called the *factor* algebra generated by x.

And we may now define Restricting Algebras in general: For B a boolean lattice the set of maps from B into B is a pointwise lattice and for each $0 < x \in B$, $\downarrow x$ is the *restricting algebra* generated by x.

Definition 2. Restricting modifier algebras are factor algebras generated by the identity function.

In addition to adjectives, PPs and relative clauses — nominal modifiers, manner adverbs constitute modifier algebras of predicates: *Max is speaking slowly* $\models$ *Max is speaking*. And they clearly combine with the boolean operators:

(4) a. He works slowly and carefully.

   b. He works slowly but not carefully.

   c. He works neither rapidly nor carefully.

So far then we have seen three distinct types of boolean structures used in models of elementary English: (1) n-place **predicate** algebras — plain vanilla complete atomic ones but with cross product structure for $n > 1$; **argument** algebras, ones that have individuals, that is, ca-free generators, and **modifier** (factor) algebras of restricting functions. And while these are all boolean algebras (lattices) they are structurally quite different.

In particular the role of boolean structure is deeper in some cases than others. For restricting functions (modifiers) the property of being restricting is defined in terms of the boolean order on their domain, but any set with a partial order admits of restricting functions, the order need not be specifically boolean (with glbs, lubs, distributive, complemented, etc.). The boolean structure evident in (4) is a reflection of how we present the modifiers, a kind of overlay.

In contrast in our major argument algebra, the generalized quantifiers, the full set of objects in our domain is *created* by boolean operations beginning with individuals. We start with n individuals and by applying the boolean operations we create 2 to the power 2^n GQs. Szabolcsi and Zwarts (1997) support that this difference shows up in several unexpected ways in what is more usually considered syntax. They cite pairs like (5a,b) and (6a,b) to show contexts where we can question and relativize when these operations apply to the GQ domain but not the manner/modifier domain:

(5) a. Which man didn't you invite?

   b. *How didn't you behave?

(6) a. Which man do you regret that I invited?

   b. *How do you regret that I behaved?

Equally in the n-place predicate lattices we find operations of much interest to linguists that have little or no analogue in the GQ or Modifier domains: valency affecting operations such as Passive, Reflexive, Causative, Applicatives come to mind. Compare:

(7) a. John opened the door   a′. $I_j((\mathbf{the(door)})(\mathbf{open}))$
     b. The door Pass(open)   b′. $\bigvee_I(I((\mathbf{the(door)})(\mathbf{open})))$

In (7b) Pass(open) is a P_1 and I ranges over individuals.

Determiners   The one other category we have considered in detail is that of Determiner (Quantifier). In some ways these behave like higher order restricting modifiers. Dets and Modifiers both take CNP denotations (properties) as arguments. A restricting modifier F picks out for each property p a subproperty $F(p) \leq p$. For F a Det function, F picks out, in effect, a set of subproperties of p, and $F(p)$ holds of a property q iff it holds of $p \cap q$, a subproperty of p. So Determiners are algebras of conservative functions, complete and atomic but not freely generated.

Another category in which conservative functions may be prominent are with quantificational adverbs: *always, often, sometimes, rarely, never*, etc. De Swart (1996) makes a good case that such quantifiers exhibit several core properties of the nominal quantifiers studied here, in particular conservativity and various monotonicity properties.

(8) a. Always/often/sometimes when Paul is tired he is in a bad mood.

     b. Always /... when Paul is tired he is tired and in a bad mood.

Much study remains to be done in this area, as we are quantifying over eventualities (or situations, or some such) which are ontologically much more complex and less well understood than simple sets of objects.

There are doubtless many other semantic functions of other categories of expression and whether they present any distinctive boolean properties is an open question here. Various modals, such as *can, must, should, may*, etc. have been studied in modal logic and certainly seem to exhibit reflexes of universal and existential quantification (van Benthem 2010). Ter Meulen (1995) argues that aspectual verbs — *begin, continue, finish*, etc. present monotonicity properties reminiscent of generalized quantifiers. The underlying partial order relation is the *part of* relation, as in *Chapter 1 is part of this book*. And if you have begun to read Chapter 1 you have begun to read this book; if you continue to read Chapter 1 you continue to read this book. So *begin* and *continue* are monotonic increasing. But *finish* is monotone decreasing: if you finish reading this book you have finished reading Chapter 1.

A last function I'll mention just to get it on the table is what I call an *indexing* function. It seems to me that locative and temporal "modifiers" sometimes serve to index — to specify evaluation "points" at which states hold or events take place. Contrast (9a,b):

(9)   a.   John is laughing in the kitchen.

      b.   John is laughing in Ben's picture.

(9a) entails *John is laughing* (*now*), but (9b) does not. Rather *in Ben's picture* serves to change the index (?possible world) at which the main clause *John is laughing* is evaluated. Similarly consider:

(10)   More people buy Toyotas every year.

Here *every* is not really quantifying over years, rather it indicates an increasing enumeration of years. We cannot naturally replace it with *some year* or *ten years*. Somehow we understand from (10) that as we enumerate years in increasing order the number of Toyota buyers increases. So *every year* here has an indexing function. Myself, I'd prefer studying indexing expressions like the two cited here in preference to positing seemingly undenotable "possible worlds".

Chapter 9

English is Inherently Sortal

In this chapter we study some as yet unexpected properties of the Noun arguments of Dets, e.g. the logical role of *poet* in *some/all/most poets* or the pair (*student, teacher*) in *more students than teachers*. To be sure the Noun argument(s) specify the conservativity domain — the set(s) of things we can limit ourselves to for purposes of the DP(s) at hand. But just how essential is this restriction? When the Det is existential (intersective) or universal (co-intersective) the Noun argument can be eliminated in favor of a neutral expression denoting the entire universe and we quantify over all the entities in the model.

Theorem 1 (Sortal Reducibility). For D of type (1,1),

    a. if D is intersective then $D(p)(q) = D(E)(p \cap q)$ since $p \cap q = E \cap (p \cap q)$, and

    b. if D is co-intersective then $D(p)(q) = D(E)(p \rightarrow q)$
       Recall: $(p \rightarrow q)$ abbreviates $(\neg p \cup q)$. So if $p - q = s - t$ then
       $\neg(p - q) = \neg(s - t)$ and $\neg(p - q) = \neg(p \cap \neg q) = \neg p \cup q = p \rightarrow q$; and
       $\neg(s - t) = (s \rightarrow t)$, so $D(E)(p \rightarrow q) = D(E)(s \rightarrow t)$.

(1) a. Some/More than ten students are vegans $\equiv$
        Some/More than ten individuals are both students & vegans

      b. All poets are vegans $\equiv$
        All individuals are s.t. if they are poets then they are vegans

The examples in (1) illustrate the theorem. The idea is that in these cases quantification over the set of students/poets can be replaced by quantification over the entire universe, compensating by using a P_1 that is built as a boolean compound of the original P_1 and the Noun, the function being intersection (meet) for intersective Dets and relative complement for co-intersective ones.

Indeed learning to effect these replacement operations occupies some time in beginning logic courses in philosophy. Generalizing we define:

Definition 1. A D of type (1,1) is *sortally reducible* iff there is a two place boolean function h such that $D(p)(q) = D(E)(h(p,q))$.

And one wonders if all D of type (1,1) are sortally reducible (the terminology derives from the use of 'sort' to refer to restrictions on the domain of quantification). The answer is a strong NO:

Theorem 2. D of type (1,1) is sortally reducible iff $D \in$ INT $\cup$ CO-INT.

A proof by (16) cases of this theorem is in Keenan (2000); a more succinct one is in Keenan (1993). To sample why the theorem holds here are the 16 boolean functions (of two arguments, given set theoretically). First define 8 such functions $h_1, \ldots, h_8$ as follows:

(2) $h_1pq = E$; $h_2pq = p$; $h_3pq = q$; $h_4pq = p \cap q$; $h_5pq = p \cap \neg q$; $h_6pq = \neg p \cap q$; $h_7pq = \neg p \cap \neg q$; $h_8pq = (p - q) \cup (q - p)$

The other eight are the functions $g_ipq = \neg(h_ipq)$. If F reduces via h_i or g_i, $i \in \{1, 2, 3, 6, 7\}$, e.g. $Fpq = F(E)(h_ipq)$, then F is constant and so both intersective and co-intersective. For $i = 3$ we use the fact that F is conservative. If F reduces via h_4 or g_4 it is non-trivially intersective and if it reduces via h_5, g_5, h_8 or g_8 it is co-intersective.

To get a feeling for these cases without going through them all consider an F that reduces via h_3. So for all p, q $Fpq = F(E)(h_3p, q) = F(E)(q)$, def $h_3, = F(\varnothing)(q)$, since p is arbitrary, $= F(\varnothing)(\varnothing)$ by conservativity. Since p, q were arbitrary, F is constant.

A second example, let F reduce via h_8. Then $Fpq = F(p)(p \cap q)$ by Cons, $= F(E)(h_8p, p \cap q) = F(E)(p - (p \cap q)) \cup (q \cap p) - p) = F(E)(p - q)$, so F is co-intersective. □

Proportionality Dets that are not int or co-int are not sortally reducible. Consider *most* (in the sense of *more than half*; if you don't like *most* used this way use *more than half* in what follows). To see that **most** is not intersective consider a model with 100 individuals, 10 of whom are poets, of which 8 daydream. Then *Most poets daydream* in that model is true, as 8 of the ten daydream. But *Most individuals are poets who daydream* is false, as only 8 of the 100 meet that condition. To see that **most** is not co-intersective consider again a 100 person model with just 10 poets, only 3 of whom daydream. Then *Most poets daydream* is false, as only 3 of the ten do. But *most individuals are either non-poets or daydream* is true, as 93 of the 100 individuals meet this condition. So **most** is not co-intersective. Ah, so suddenly the proportionality Dets, largely ignored up to now, become of interest. They show that the apparent restriction the Noun argument places on DPs is not eliminable in the way it is with standard logical quantifiers. And something stronger can be said: typical examples of conjunctions and disjunctions of int with co-int Dets fail to be either int or co-int. E.g. *some and not all* does not denote an int function (because of the *not all* part) and does not denote a co-int function (because of the *some* part).

And one is led to wonder just what portion of the conservative functions actually are intersective or co-intersective? In a model E, INT_E and CO-INT_E have 2 to the power 2^n elements with a two element overlap, whereas CONS_E has 2 to the power 3^n elements. So if $|E| = 2$, $|\text{INT}_E \cup \text{CO-INT}_E| = 16 + 16 - 2 = 30$ and $|\text{CONS}_E| = 512$. If $|E| = 3$, $|\text{INT}_E \cup \text{CO-INT}_E| = 510$ and $|\text{CONS}_E| = 2^{27} \approx 134,000,000$ so the conservative non-(co-)int functions massively outnumber those that are (co-)intersective. So nearly all DPs of the form Det+N use the Noun restriction non-trivially.

Moreover, from Theorem 2, this non-trivial usage is due to closing $\text{INT}_E \cup \text{CO-INT}_E$ under the boolean operations. Note that this closure includes all the proportionality functions. The elements of the set we are closing are sortally reducible.

Also many of the functions added in by the closure are not themselves proportional. *The ten* is not, nor are partitives like *more than two of the ten*.

Prop$_E$, Proportionality Det Denotations   The F in Prop$_E$ recall are those that map pairs p, q and u, v to the same value if the proportion of p's that are q's is the same as the proportion of u's that are v's. In computing the proportion of p's that are q's, $|p \cap q|/|p|$, we assume $|p|$ is finite and $|p| > 0$.

One sees easily that if F, G of type (1,1) are proportional so are $\neg F$ and $(F \wedge G)$ and $(F \vee G)$. E.g. [*more than a quarter but less than half*] of *American teenagers are overweight* just depends on the proportion of American teenagers that are overweight. And since proportional Dets are conservative, assuming E finite we have that PROP_E is a complete sublattice of CONS_E, so it is atomic (An atom in PROP_E is given by taking the glb of the elements of PROP_E that lie above a fixed atom in CONS_E). So its cardinality is 2^k where $k = |\text{Atom}(\text{PROP}_E)|$.

And one shows that the atoms of PROP_E are determined by the *reduced* fractions k/m, where $0 \leq k \leq m$, $0 < m \leq n$, $n = |E|$. They are:

Theorem 3. The atoms of PROP_E are the functions $\text{Ex}(k/m)$, for k, m as above, given by: $\text{Ex}(k/m)(p)(q) = 1$ iff $|(p \cap q)| = (k/m) \cdot |p|$.

Different reduced fractions meeting the conditions determine different atoms, so to count the atoms we must count the reduced fractions whose denominators lie between 1 and $n = |E|$. Our atom count uses Euler's *totient* function φ, where $\varphi(m)$ is the cardinality of the set of positive integers p whose greatest common divisor with m is 1 (that is, p is prime relative to m). Then:

Theorem 4. $|\text{PROP}_E| = 2^{\text{RF}(n)}$, $n = |E|$ and $\text{RF}(n)$ is defined by:

$$\text{RF}(n) = 1 + \sum_{1 \leq k \leq n} \varphi(k).$$

A more elegant recursive definition is:

$$\text{RF}(1) = 2 \text{ and for all } n > 0, \text{RF}(n+1) = \text{RF}(n) + \varphi(n+1).$$

The 1 in the theorem represents the case where $|p \cap q| = 0$, and $0/m$ is not a reduced fraction so it is counted separately. If $|E| = 1$ there is only one reduced fraction, $1/1 = 1$. So $\text{RF}(1) = 2$, counting the 0 case. For $|E| = 2$ there are two reduced fractions $\frac{1}{2}$ and $\frac{2}{2}$, but the latter has already been counted as $1/1$, so we just add in one more, and $\text{RF}(2) = 3$. The new fractions we add in as we move from $|E| = n$ to $|E| = n + 1$ are just the fractions k/m where k and m have no divisor greater than 1 (since if they do that fraction has already been counted earlier). Thus we just add in at line $n+1$ the k/m where k is relatively prime to m, which is what Euler's function computes. (Note: evaluating $\varphi(n)$ uses the prime factors of n, though not how many times they occur.) Here are a few values of $\text{RF}(n)$ for low n:

(3) | n | 3 | 5 | 6 | 8 | 10 | 15 |
      |----------|---|----|----|----|----|----|
      | $\text{RF}(n)$ | 5 | 11 | 13 | 23 | 33 | 74 |

By way of comparison in a model with just 5 entities there are $2^{5+1} = 64$ cardinal functions and $2^{11} = 2,048$ proportional ones. Note that we have only considered proportionality Dets that are properly "mathematically invariant", a notion we define later in terms of permutation invariance. So the relevant comparison here is with CARD_E not INT_E. *What per cent of?* seems like a reasonable interrogative Det (which of course forms questions, like the other interrogative Dets we have considered) an issue not pursued here.

A "traditional" semantic role of Nouns   In natural languages n-place predicates impose semantic *selection restrictions* on their DP arguments. The Ss in (4a) are natural, those in (4b) anamolous (marked #):

(4) a.   Some/Every student laughed at my joke.

 b. #Some/#Every ceiling laughed at my joke.

The anomaly in (4b) is due to the fact that ceilings aren't the kinds of things
that can laugh. If I say that "X laughed at my joke" and you didn't hear
'X' you can still infer that X is animate, indeed moderately well placed on
some chain of being hierarchy — humans laugh, maybe dolphins, but beetles,
ceilings and numbers definitely don't laugh.

Now the selection restrictions of a DP are expressed by the Noun, not
the Det. Changing the Dets in (4b) to e.g. *exactly one* or *most* preserves
the anomaly. Making this change in (4a) preserves its naturalness. In DPs
built from Dets of type $((1,1),1)$ normally both Noun arguments satisfy the
selection restrictions of the predicate: *Fewer girls than boys laughed*, but
#*Fewer girls than ceilings laughed* and #*Fewer ceilings than girls laughed*. So
it is the Noun properties of a DP that determine whether the DP satisfies the
selection restrictions imposed by the predicate. (Rather different properties,
such as monotonicity and licensing of negative polarity items, are determined
by the Det.)

Chapter 10

Logical Operators and Structure Maps

We are initially concerned here with how to characterize the "logical" character of quantifiers like *some, all, most, most but not all*, etc. compared with "non-logical" Dets such as *John's two* and *more male than female*. The issue may seem "philosophical" but we see that theoretical linguistics builds in a kind of logical/non-logical distinction into syntactic theories. We conclude by extending our notion of semantic invariance to syntax and speculate concerning the relation between syntactic and semantic invariants.

We will see that in "large" categories such as type (1,1) Dets, logicality constrains the possible denotations that lexical items in that category can have (Keenan 1986), so it determines a non-trivial linguistic generalization. In "small" categories lexical items denote more freely, but the subset of "logical" denotations is inherently bounded, in contrast with large categories where that set scales (exponentially) with the size of the universe.

On the sizes of semantic sets   Of course $\{0, 1\}$ has cardinality 2, and $|E|$ can be any $n > 0$. In this chapter we assume $|E| \geq 2$ and finite unless stated otherwise. (Most of our claims hold when $|E|$ is infinite, but many figures, such as 2^n and 2^{2n} collapse to the same infinite cardinal when n is infinite so the comparison is not revealing.) Our other semantic sets are function spaces, $[A \to B]$, which have cardinality $|B|^{|A|}$ when the full set of functions is considered. Usually however our functions of interest are ones that satisfy certain conditions. E.g. extensional adjective denotations are *restricting* maps from $P(E)$ to $P(E)$: $F(p) \subseteq p$.

The function spaces we study all are complete atomic boolean lattices (algebras). The facts in (1), partly given earlier, are used often in computing the force of logicality constraints on boolean sets:

(1)   Let $(B, \leq)$ be a complete and atomic (ca) boolean lattice. Then

    a. $(B, \leq) \simeq P(\mathrm{Atom}(B))$, so $|B| = 2^{|\mathrm{Atom}(B)|}$, and
    b. for all $q \in B$, $q = \bigvee\{\alpha \in \mathrm{Atom}(B) \,|\, \alpha \leq q\}$, and
    c. if $(B', \leq')$ is a complete sublattice of $(B, \leq)$ it is also atomic.
        $(\beta \in \mathrm{Atom}(B')$ iff For some $\alpha \in \mathrm{Atom}(B)$, $\beta = \bigwedge\{x \in B' \,|\, \alpha \leq x\})$

So typically to count the sets which satisfy various linguistic constraints thereby observing how strong they are, we show that they are complete sublattices of some big lattice and we count the atoms.

    E.g. type (1,1) is given by: $[P(E) \to [P(E) \to \{0, 1\}]]$ which has cardinality $2^{|P(E) \times P(E)|}$, and $|P(E) \times P(E)| = 2^n \times 2^n = 4^n$. So there are 2 to the 4^n type (1,1) functions. Only 2 to the 3^n are conservative. We exhibit the atoms of CONS and show they number 3^n.

Fact 1. The atoms of CONS are the functions $f_{p,q}$, with $q \subseteq p \subseteq E$ defined by: $f_{p,q}(s)(t) = 1$ iff $s = p$ and $s \cap t = q$. (So $f_{p,q}$ maps only p to a non-zero quantifier, and the value $f(p)$ assigns to a property t depends on the intersection of t with p.)

Fact 2. The number of $f_{p,q}$ is the number of pairs (p, q) with $q \subseteq p$, the same as $|[\mathrm{Atom}(P(E)) \to \{1, 2, 3\}]|$. Given a g in this set we uniquely construct q as $\bigvee\{\alpha \in \mathrm{Atom}(P(E)) \,|\, g(\alpha) = 1\}$ and p as $\bigvee\{\alpha \in \mathrm{Atom}(P(E)) \,|\, g(\alpha) = 1 \text{ or } g(\alpha) = 2\}$. Clearly $q \subseteq p$. Further, each pair (p, q) with $q \subseteq p$ uniquely determines such a function, as indicated below:

$$g_{p,q}(\alpha) = \begin{cases} 1 & \text{if } \alpha \in p \cap q \; (= q) \\ 2 & \text{if } \alpha \in p - q \\ 3 & \text{if } \alpha \in \neg p \end{cases}$$

    So there are 3^n atoms in CONS so the number of conservative functions is 2 raised to that power. (These computations do not require n to be finite.) Now we turn to the primary topic of interest:

Logicality   To introduce the idea suppose that $(B, \leq_B)$ and $(D, \leq_D)$ are two boolean lattices. Then, standardly, they are isomorphic ($\simeq$) iff there is a bijection h from (say) B to D such that for all $x, y \in B$, $x \leq_B y$ iff $h(x) \leq_D h(y)$. (It follows that h^{-1} is an isomorphism from $(D, \leq_D)$ to $(B, \leq_B)$.) The notion of isomorphism of mathematical structures of course applies to arbitrary structures — sets with various functions and relations defined on them satisfying certain conditions, but here we limit ourselves to the relevant boolean structures.

An isomorphism h from a mathematical structure to itself is called an *automorphism*. In particular an automorphism of a boolean lattice $(B, \leq)$ is a bijection h from B to B which does not change the $\leq_B$ relation: $x \leq y$ iff $h(x) \leq h(y)$, and we say that h *fixes* $\leq$ and write $h(\leq) = \leq$. This implies that h fixes all functions and relations defined in terms of $\leq$. I.e. $h(1_B) = 1_B$ and $h(0_B) = 0_B$; $h(\wedge_B) = \wedge_B$ and ditto for $\vee$ and $\neg$. The former means that $(x \wedge y) = z$ iff $(h(x) \wedge h(y)) = h(z)$, that is, $h(x \wedge y) = h(x) \wedge h(y)$. In fact h fixes arbitrary glbs (and lubs): $h(\bigwedge_{j \in J} x_j) = \bigwedge_{j \in J} h(x_j)$, so h *commutes* with (arbitrary) meets and joins.

The fact that h provably fixes the zero and unit of a BL means that the only automorphism of the $\{0, 1\}$ lattice is the identity map (the other bijection from $\{0, 1\}$ to itself is the $\neg$ function, which does not fix $\leq$; $0 \leq 1$ but $\neg 0 \nleq \neg 1$. Later we consider some philosophical issues concerning logical constants. Here we note that commonly logicians consider **truth** as a fundamental notion, and hence a structure preserving function h (e.g. an isomorphism) on the truth value lattice (algebra) must preserve truth, so h must map **truth** to **truth** (Westerståhl 1985).

Structure (preserving) maps The perspective we take here is slightly different — structure (preserving) maps (*automorphisms*; see below for a general definition) respect the *structure* on sets, not the identity of their elements. We require simply full preservation of the truth relation, $\leq$: $(p \to q)$ iff $(h(p) \to h(q))$, i.e. for all $x, y \in \{0, 1\}$, $x \leq y$ iff $h(x) \leq h(y)$ and we say that h *fixes* $\leq$, i.e. $h(\leq) = \leq$, as noted above. It follows, as in all boolean lattices, that it maps the unit to the unit and the zero to the zero.

It would perhaps have been more in keeping with our perspective if we had allowed the objects in the domain of the truth value lattice to vary from model to model, as we do with the universe. Sometimes it would be $\{0, 1\}$, sometimes $\{\mathbf{F}, \mathbf{T}\}$, etc. But it would always be a boolean lattice, and since we are not (here) interested in issues concerning many valued logics we would keep it to a two element set. But since any two finite boolean lattices of the same cardinality are isomorphic, it would be isomorphic to $(\{0, 1\}, \leq)$, the standard choice, which we stick with as it is easier to refer to 0 and 1 rather than "the zero of the truth value lattice" and "the unit …", etc.

Returning to automorphisms h of boolean lattices $(B, \leq)$ we note that for any $K \subseteq B$, $h(K)$, often noted $h[K]$, is $\{h(x) \mid x \in K\}$. So $h(B) = B$ (since h is onto) and $h(\varnothing) = \varnothing$. So any automorphism of B fixes $\{1_B, 0_B\}$. More interesting for us is $h(\text{Atom}(B)) = \text{Atom}(B)$, even if B is not atomic (atoms are defined in terms of $\leq$).

Models again We have been taking $(E, \{0, 1\})$ as the semantic primitives of our (assumed) English, where $\{0, 1\}$ is the minimal boolean lattice but E is just a random non-empty set, with no general designated functions or relations. So an automorphism of E is just a permutation of E. It preserves vacuously "all" the designated functions and relations on E. To avoid confusion we assume that $\{0, 1\}$ is disjoint from E, and that E has at least two elements (so $P(E)$ has at least four). Later we dispense with E but we do need a ca boolean lattice P to play the role of $P(E)$. For the moment for reasons of familiarity we stay with E and $P(E)$. But note that once we interpret proper nouns and DPs like *the poet* as individuals, I_bs, they are maps from $P(E)$ into $\{0, 1\}$ so we have no expressions which denote in E. The closest we come is common nouns (*poet*, etc.) which denote subsets of E.

Now we note that for B a ca boolean lattice — we often omit reference to the relation $\leq$ — the automorphisms of B are given by the permutations of $\text{Atom}(B)$. If h is such a permutation then, using α as a variable ranging over $\text{Atom}(B)$, for all $q \in B$, $h(q) = h(\bigvee_{\alpha \leq q} \alpha) = \bigvee_{\alpha \leq q} h(\alpha)$. In the case where $B = P(E)$, a permutation g of E extends to a permutation h_g of $\text{Atom}(P(E))$ by $h_g(\{b\}) = \{g(b)\}$. The map sending each g to h_g is bijective, so the automorphisms of $P(E)$ correspond one for one to the permutations of E.

Denotation sets for categories of expressions $\text{TH}(E, 2)$, the *Type Hierarchy* generated by $\{E, \{0, 1\}\}$, is the least set containing E and $2 = \{0, 1\}$ and closed under the formation of function spaces, that is, $[A \to B]$ is in it whenever A and B are. Elements of $\text{TH}(E, 2)$ are called *types*. The *t-types* are$_{\text{def}}$ $\{0, 1\}$ and all those of the form $[A \to B]$ where B is a t-type. If C is a category of expression, $\text{Den}_E C$, the set of possible denotations of expressions of category C relative to E and 2, is a subset of a type in $\text{TH}(E, 2)$.

Definition 1. For each (disjoint) pair $(E, 2)$,

  a. A *basic automorphism* is a permutation h of $E \cup 2$ s.t. $h \restriction E$ is a permutation of E and $h \restriction 2$ is the identity map on $\{0, 1\}$.
  b. An *automorphism* of $\text{TH}(E, 2)$ is an extension of a basic automorphism h to $[A \to B]$ as follows: if $h \restriction A$ is an automorphism of A and similarly for B then h extends to $[A \to B]$ by setting $h(f)(h(x)) = h(f(x))$. NB: each y in A is an $h(x)$ for exactly one x in A since h is bijection of A.

c. An object **d** in a type is *automorphism invariant* (AI) iff for all automorphisms h, $h(\mathbf{d}) = \mathbf{d}$. An expression d is AI iff all its possible denotations in an arbitrary model $((E, 2), m)$ are AI.

To get a grip on what being AI means let us first consider some examples, and then turn to some of its more general logical and linguistic properties including one generalization of the notion. We note that as an automorphism is given by a permutation of E, many authors write PI for *permutation invariance* where we write AI. Now let us see first that given $E, 2$ *all* is AI. Then we prove two general and very useful properties of AI-ness. Observe:

Lemma 1. For $p, q \in P(E)$, all automorphisms h, $p \subseteq q$ iff $h(p) \subseteq h(q)$.

Proof. $\Rightarrow$ Let $p \subseteq q$. Then if $b \in h(p)$ then $b = h(x)$ some $x \in p$, so $b = h(x)$ for some $x \in q$, so $b \in h(q)$, whence $h(p) \subseteq h(q)$.
$\quad\Leftarrow$ Let $b \in p$, so $h(b) \in h(p) \subseteq h(q)$, so $b \in q$, whence $p \subseteq q$. $\qquad\square$

Theorem 1. all is AI.

Proof. Let $p, q \subseteq E$, h an automorphism. Then

$$
\begin{aligned}
h(\mathbf{all})(hp)(hq) = 1 \quad &\text{iff } h(\mathbf{all}(p)(q)) = 1 &&\text{def } h \\
&\text{iff } \mathbf{all}(p)(q) = 1 &&h = \mathrm{id} \upharpoonright \{0, 1\} \\
&\text{iff } p \subseteq q \\
&\text{iff } h(p) \subseteq h(q) \\
&\text{iff } \mathbf{all}(h(p))(h(q)) = 1
\end{aligned}
$$

Thus $h(\mathbf{all}) = \mathbf{all}$ as they take the same values at all properties. $\qquad\square$

Theorem 2. For f in a functional type $[A \to B]$, f is AI iff for all automorphisms h, $h \circ f = f \circ h$, i.e. h commutes with f.

Proof. $\Rightarrow$ Let f be AI. Then $h(f(a)) = h(f)(h(a))$, def h, $= f(h(a))$, $hf = f$, so h commutes with f.
$\quad\Leftarrow$ Let h commute with f. Then $h(f)(h(a)) = h(f(a))$, def h, $= f(h(a))$, h commutes with f. Thus $h(f) = f$ as they take the same value at all arguments $h(a)$ — (h is onto, recall). $\qquad\square$

Theorem 3. For h an automorphism of $\mathrm{TH}(E, 2)$ and B a t-type (and thus a ca BL) $h \upharpoonright B$ is a boolean automorphism of $(B, \leq_B)$. That is, $h \upharpoonright B$ is a permutation of B which fixes $\leq_B$.

Proof sketch. case 1: $B = \{0, 1\}$. There are only two permutations of $\{0, 1\}$ and if $h(0) = 1$ and $h(1) = 0$ then we have $0 \leq 1$ but $h(0) \not\leq h(1)$ so this h does not preserve $\leq$. The identity map trivially does.

case 2: $B = [X \to Y]$ where $h \upharpoonright X$ is a permutation of X and Y is a t-type with $h \upharpoonright Y$ a boolean automorphism of $(Y, \leq_Y)$. Then $f \leq g$ iff for all $x \in X$, $f(x) \leq g(x)$, iff $h(fx) \leq h(gx)$, f fixes $\leq_Y$, $hf(hx) \leq hg(hx)$, def h, iff $h(f) \leq h(g)$, since all elements of X are $h(x)$ for some x as h is onto. $\square$

Some classes of AI functions   For each category C of expression we have discussed, $\mathrm{Den}_E\, C$ is AI. E.g. if R is a k-place relation over E and so a possible P_k denotation then so is $h(R)$, all automorphisms h. This is hardly surprising given how automorphisms of $\mathrm{TH}(E, 2)$ are defined. Perhaps slightly less obvious is that the subsets of the $\mathrm{Den}_E\, C$ such as CARD, INT, CO-INT, PROP, CONS, type (1), type (1,1), etc. are each also AI. E.g. if F is a conservative function of type (1,1) so is $h(f)$. If f is a restricting adjectival function so is $h(f)$, etc. In this sense the defining conditions for semantic subcategories we have discussed are AI conditions.

One very slightly surprising case are the individuals $\{I_b \,|\, b \in E\}$. As long as $|E| > 1$ no individual is itself AI. But automorphisms map individuals to individuals: $h(I_b) = I_{h(b)}$. So the set of individuals — the property of being an individual — is AI.

Consider first the denotation sets (given E) of P_ns, the n-place predicates. For P_0 each element of $\{0, 1\}$ is fixed by the identity map and thus by "all" automorphisms as there is only one. For P_1s, $P(E)$ has just two AI elements, E and $\varnothing$. Any proper non-empty subset K of E has an element x and lacks an element y. The map h which interchanges x and y and fixes everything else extends to an automorphism of $P(E)$ but does not fix K, as it replaces x by y, so $h(K) \neq K$. At the P_2 level an interesting pattern begins to emerge. The unit $E \times E$ and the $\varnothing$ set of ordered pairs are AI, but so also is the identity relation $\{\langle x, x \rangle \,|\, x \in E\}$ and its complement, $\{\langle x, y \rangle \,|\, x \neq y\}$. No other binary relation is AI. Note that so far the denotation set for an AI P_n is a ca boolean lattice. In fact this holds for all P_n.

Given E, for D a denotation set of a category of expression, write $\mathrm{AI}(D)$ for the set of AI elements of D relative to E and $\{0, 1\}$. Then

Theorem 4. $|\,\mathrm{AI}(\mathrm{Den}_E(P_k))| = 2^{|\mathrm{EQ}(k)|}$ (provided $|E| \geq k$), where $\mathrm{EQ}(k)$ is the number of equivalence relations over a k element set. (See Westerståhl 1985 for an explicit calculation of $\mathrm{EQ}(k)$.) And as expected from the cardinality, each $\mathrm{AI}(\mathrm{Den}_E(P_k))$ is a complete atomic sublattice of $P(E^k)$ with $\mathrm{EQ}(k)$ atoms.

For example for $k = 3$ there are five equivalence relations: one where all the coordinates are the same, $\{\langle x, x, x \rangle \mid x \in E\}$, one where they are all different, and three where just two are the same: $\langle x, x, y \rangle, \langle x, y, x \rangle$ and $\langle y, x, x \rangle$. So the number of AI ternary relations is $2^5 = 32$ (given E with at least three, and perhaps infinitely many, elements).

Two remarks on AI predicates   First, a linguistic speculation: Surely all natural languages have a way of identifying individuals characterized differently. *That person is John, My employer is my brother, The maid is the one who stole the jewels.* So the equals P_2, $=$, is grammaticized (by a lexical item or a designated grammatical construction). Recall: only about half the world's languages have an overt copula *is/be*. But I know of no language which has a three place equals, holding of $\langle a, b, c \rangle$ iff $a = b = c$. Nor do we find grammaticized $=_{1,3}$, which holds of $\langle a, b, c \rangle$ iff $a = c$ but b doesn't, etc.

Second, AI-ness (Keenan 2001) plays a role in relating truth and reference. Putnam (1981) observes that we can change reference in the sense of changing the extension of predicates without changing the truth of sentences, building on the fact that isomorphic models make the same sentences true. But Putnam slightly overstates the claim in saying that we can choose a predicate whose denotation is not $\varnothing$ or universal, then find an automorphism which interchanges some x in the extension of the predicate with a y not in it preserving truth.

Putnam's claim holds when the predicate is a P_1. But the empty/universal dichotomy is insufficient for P_ks, $k > 1$. For example given $E = \{a, b\}$ with just one P_2 R denoting $\{\langle a, b \rangle, \langle b, a \rangle\}$, the three Ss below are all false, but any change in the extension of the predicate makes one of them true:

$$\exists x (xRx) \qquad\qquad \forall x \forall y \neg (xRy) \qquad\qquad \exists x \exists y (xRy \,\&\, \neg yRx)$$

So as stated Putnam's argument fails, but the basic idea is right. The correct conditioning factor is not empty/universal but AI-ness. That is, if the denotation of a predicate is not AI we can change its extension preserving truth, but not if the denotation is AI.

Following up this philosophical observation let us hypothesize, consonant with recent work by logicians (Westerståhl 1985; van Benthem 1989a) though at variance with Etchemendy (1990), that:

Thesis. Logical constants always denote AI objects

This (weak) Thesis raises three issues. First, consider:
Hypothesis A An expression d is

a. *logical* iff for all models $(E, 2, m)$, $m(d)$ is AI,
b. *(locally) constant* iff for all models $((E, 2), m)$ and $((E, 2), m')$, $m(d) = m'(d)$.

Theorem 5 (Keenan and Stavi 1986). d is constant $\Rightarrow d$ is logical.

Proof. Let $(E, 2, m)$ arbitrary, h a permutation of E. Then $h(m(d)) = (h \circ m)(d) = m(d)$ by (local) constancy, so $m(d)$ is AI. □

We have assumed (correctly) that models are closed under automorphisms, that is, $h \circ m$ is a possible interpreting function.

Westerståhl and van Benthem would strengthen our definitions above by replacing AI with *isomorphism invariant* (ISOM), called QUANT in van Benthem (1986, Ch 2), where:

Definition 2. For D a functional mapping each non-empty E to a function from k-tuples of properties to truth values, D is ISOM iff for all E, E' all bijections $h : E \rightarrow E'$, $D_E p_1, \ldots, p_k = D_{h(E)} h(p_1), \ldots, h(p_k)$. This just says $D_{h(E)} = h(D_E)$, where h extends to TH$(E, 2)$ as before.

Now h being AI is just the special case of ISOM in which $h(E) = E$. (But note that this view of logicality forces us to shift our concept of a Det from something that is simply a member of a $\tau \in$ TH$(E, 2)$ for some E to a functional which surveys the proper class of all non-empty sets E.) And we may generalize the hypothesis above to:

Hypothesis B An expression d which denotes a functional as above

a. is *logical* iff $m(d)$ is ISOM, all models $(E, 2, m)$, and
b. is *(globally) constant* iff for all models $(E, 2, m)$ and $(E', 2, m')$ and all bijections $h : E \rightarrow E'$, $h((m(d))_E) = (m'(d))_{E'}$.

An advantage of the generalization (Keenan and Westerståhl 1997) is that a functional like D given by $D_E pq = 1$ iff $7 \in E$ is not ISOM but is AI (all E; D_E is either the unit $\mathbf{1}$ or the $\mathbf{0}$ according as $7 \in E$ or not). And a d that denotes D is not, pretheoretically, constant.

But AI-ness is independently interesting. It enables us to see that logicality $\not\Rightarrow$ constancy. Consider Dets such as *about/approximately* 50. They are "precisely vague". We might account for this by interpreting *about 50* sometimes as *49*, sometimes as *51*, etc. without being able to say precisely what delimits our interpretative freedom. Still, *about 50* would be AI in that its denotation, while not constant across models (even with the same universe), is always AI. But *about 50* is not ISOM: no isomorphism maps **exactly 49** to **exactly 51**. *Several* and *nearly 50* are similarly vague.

An apparent appeal of ISOM is its global nature: it quantifies over all universes, whereas being AI is local, decided universe by universe. However AI-ness is equivalent to a global notion, but one that is properly weaker than ISOM. Namely,

Definition 3. A functional Q of type $(1,1)$ is *weakly* ISOM iff for all universes E, E' all bijections $g, h : E \to E'$, $g(Q_E) = h(Q_E)$.

Theorem 6. Q is AI iff Q is weakly ISOM.

That is, all automorphisms agree on Q_E iff all isomorphisms to the same universe agree on Q_E. Note that if all auts agree on some X then they map X to itself since $\mathrm{id}_{E,2}$, the identity map on E and $\{0,1\}$, an automorphism that extends to id_τ, each $\tau \in \mathrm{TH}(E, 2)$.

Proof. $\Leftarrow$ is obvious, setting $E' = E$. For $\Rightarrow$, assume Q_E is AI and that $g(Q_E) \neq h(Q_E)$. Then for any bijection $f : E' \to E$, $f(g(Q_E)) \neq f(h(Q_E))$, so $(f \circ g)(Q_E) \neq (f \circ h)(Q_E)$. But this is false, as $f \circ g$ and $f \circ h$ are automorphisms, so they take the same value at Q_E assumed AI. $\qquad\square$

So ISOM is of some value in characterizing logical constants — it rules out the D whose value at E varied with whether 7 was in E or not. But it allows "nonconstancy" conditioned by ISOM factors. E.g. let $Q_E =$ **some** if $|E| = 7n$, n finite, otherwise $Q_E =$ **all**. It seems to me then that ISOM buys us little that is not already included in the AI package. And in practice the "logical" properties of objects we study derive from AI not the additional coverage afforded by ISOM.

Even in topic specific mathematical theories — set theory, metric spaces, Euclidean geometry, etc. *individual objects* are only identifiable in so far as the primitive relations of the field imply their existence and uniqueness. Consider for example the (finite rooted) trees that linguists use. Such a tree is a pair (N, D), where N is a finite non-empty set (of *nodes*) and D is a reflexive partial order, *dominates,* which satisfies two further axioms: (1) there is a node which dominates all nodes and (2) whenever x and y are dominance independent (neither dominates the other) then they don't dominate anything in common.

Axiom (1) says there is at least one node which dominates every node. Uniqueness follows: if both x and x' dominate everything then each dominates the other so they are identical by antisymmetry (nDm and $mDn \Rightarrow n = m$). So we can refer to this unique object as **the root**. Similarly in boolean lattices by axiom the domain has least and greatest elements, by theorem there is just one of each, so we can baptize them, '0' and '1'. In set theory by axiom there is a set with no members, by theorem there is just one. Call it '∅', etc.

So it seems even more natural in modeling natural languages, which are *general purpose* — we can use them to talk about anything, we can't assume the existence of specific objects independent of our study. So it is for epistemological not logical reasons that we cannot assume the number 7 (or the Buddha) to be in or not in our domain. It might help to reflect on just how "structure preserving maps" can identify particular objects. A mathematical structure **A** is a set A (or several sets) on which are defined relations (which may be functions) with specified properties — these relations constitute the "structure" of A. And a fully structure (preserving) map is a permutation of A which fixes the relations, i.e. doesn't change which tuples of objects are in the relations. And if it happens that the conditions the relations in a structure must satisfy are strong enough to imply the existence of a unique element with certain properties then all structure maps must map it to itself — that object is *structurally definable*.

To see neutrally how structure implies (unique) existence consider the class of unordered rooted trees with just four nodes depicted below (We read *down* from x to y to say that x strictly dominates y):

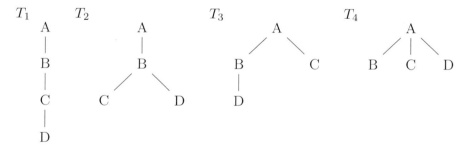

No two of these trees are isomorphic and any 4-node unordered rooted tree is isomorphic to one of them.

Now in all the trees the *root* node is dominance definable as that (sole) node that dominates all nodes. In T_4 only the root node A is dominance definable — B, C, and D are indistinguishable in terms of dominance: A strictly dominates each of them and each strictly dominates nothing. In T_2 A and B are dominance definable: B the unique node strictly dominated by the root. But C and D are dominance indistinguishable. In T_1 each node is definable: B the unique node which the root immediately dominates (strictly dominates, with no node z which the root strictly dominates and which strictly dominates it). C is the unique node which B strictly dominates and D the unique node that strictly dominates no nodes.

Interestingly in T_3 each node is dominance definable. Thus in T_1 and T_3 the only automorphism (structure map) is the identity map. For T_2 there are two automorphisms: the identity map and the map that transposes C and D and fixes the other nodes. In T_4 there are six automorphisms: the permutations of {B,C,D} which fix A.

A deeper issue raised by Westerståhl (1985) is just why words like *all*, *not*, and *and* are constrained to be constantly interpreted but ones like *blue* are not. There is a possible confusion here in this way of putting the question. Obviously there is no inherent link between a phonemic or orthographic sequence and fixed interpretation. /si/ means *and* in Malagasy but *yes* in Spanish, /nain/ means 9 in English but *no* in German, etc. The issue is that the denotations of *all*, *not*, etc. are AI (or ISOM) while that of *blue* is not. There is no more reason for this than that *dog* doesn't mean *cat* and vice versa.

But Westerståhl's serious question remains: why should we take invariance under automorphisms or isomorphisms as a necessary condition for an object to be "logical"? Westerståhl (1985) has two suggestions. One, isomorphisms preserve the truth of sentences (isomorphic models are elementarily equivalent), and truth is logically fundamental. And two, the items we treat pretheoretically as logical constants play a central role in making inferences, ones which quantify over the "content" words: if *all* As *are* Bs *and all* Bs *are* Cs then *all* As *are* Cs. Replacing *all* by *some* here invalidates the inference. But we have infinitely many choices for A, B, and C and the inference is valid in all cases. For a rich and convincing variety of such inference patterns see van Benthem (1986, Ch 2).

The somewhat more boolean perspective we have taken on natural language semantics here may add slightly to this picture. We have taken sentence denotations to be a relational structure, the boolean lattice $(\{0,1\}, \leq)$. An automorphism of this structure is a bijection of $\{0,1\}$ which fixes the $\leq$ relation, the choice of objects 0,1 in the field of the relation is irrelevant.

And $\leq$ is basically the truth table for the conditional $(p \rightarrow q)$, the primary inferencing connective. The AI elements of $[E \rightarrow \{0,1\}]$, of $[E \rightarrow [E \rightarrow \{0,1\}]]]$, etc. follow from this as do $[[E \rightarrow \{0,1\}] \rightarrow \{0,1\}]$, etc. So on the view we are suggesting the root of AI-ness is preservation of the boolean relation, other operators like *and*, *or*, *not*, etc. are characterized in terms of it (as glbs, lubs, etc.).

Lastly, what about the converse of Theorem 5? Does denoting an ISOM (or AI) object in all models suffice to make an expression a logical one? Not likely, as the set theoretical relations and operations are AI. And while simple ones, like $\subseteq$ and $\cap$ may seem "logical", more esoteric claims — e.g. every set can be well ordered — seem to be mathematical truths of set theory but not clearly "logical" ones any more than general truths of topology are. We should note though (Keenan 2001, Theorem 22) that being "logical" is a logical property:

Theorem 7. For each E, each $\tau \in \mathrm{TH}(E, 2)$, the function Log which maps each x in τ to 1 iff x is AI is itself AI.

Two linguistic generalizations re logicality   Notice $|\mathrm{AI}(\mathrm{Den}_E\, P_k)|$ is bounded as a function of the arity k of the predicates. In contrast consider the Det denotation **more than m** where $m > |E| = n$. All these functions map all pairs p, q to 0. But if we add more elements to E then **more than m** becomes non-trivial for some $m > n$, so the number of AI elements scales with $|E|$, i.e. increases with the size of E.

Let us compare denotation sets $\mathrm{Den}_E(C)$ for familiar categories C in terms of how fast they grow. As they are all ca BLs, $|\mathrm{Den}_E(C)| = 2^m$, m the number of its atoms. Call C *small* if m is a polynomial function of $n = |E|$ and call C *large* if m is exponential in n. So the k-place predicates are a small category as their exponent is n^k, polynomial in n. But $\mathrm{Den}_E\, \mathrm{DP} = [P(E) \to \{0, 1\}]$ has size 2^k for $k = 2^n$, exponential in n, not polynomial. And for the type (1,1) Dets, $k = 4^n$; for the CONS subset it is 3^n, and for INT and CO-INT it is 2^n, always exponential in n. So Dets and DPs are *large* categories as their denotation sets scale hyperexponentially with $|E|$. Generalizing:

Gen 1. For D_E a denotation set for a category C of expression, if $\mathrm{Den}_E(C)$ is small then $\mathrm{AI}(\mathrm{Den}_E(C))$ is bounded independently of $|E|$; if $\mathrm{Den}_E(C)$ is large $\mathrm{AI}(\mathrm{Den}_E(C))$ scales with $|E|$.

As we have seen, Gen 1 holds for k-place predicates. The large category of primary interest here is Det_1 whose denotation set, CONS, is hyperexponential in n. And we have:

Theorem 8. $|\mathrm{AI}(\mathrm{CONS}_E)| = 2^{(n+1)(n+2)/2}$ (E finite)

The atoms of CONS are given by expressions of the form: *exactly k of the m* for $0 \leq k \leq m \leq |E|$. They denote the functions $f_{k,m}(s)(t) = 1$ iff $|s| = m$ and $|s \cap t| = k$. We might note as well that the atoms of the intersective

Dets of type (1,1) are just the functions *exactly k* which map p, q to 1 iff $|p \cap q| = k$. In fact:

Theorem 9. For E finite, $\mathrm{AI}(\mathrm{INT}_E) = \mathrm{CARD}_E$, of cardinality $2^{|E|+1}$.

The assumption that E is finite is required here as the proof uses the fact that if $|p| = |q|$ then there is an automorphism h that interchanges them. But such an h must also map $\neg p$ to $\neg q$, which could fail if E were infinite. E.g. for $E = \mathbb{N}$ we have that $|\mathrm{EVEN}| = |\mathbb{N}|$ but no automorphism can map EVEN to $\mathbb{N}$ (or any cofinite subset of it) as it would have to map ODD to $\varnothing$ (or some finite set), which cannot be.

DP is also large and most DPs of the form Det+N are not AI. E.g. *all poets* is not: Let $E = \{a, b, c\}$ and **poet** $= \{a, b\}$. Let h a permutation of E which transposes b and c and fixes everything else. Then $h(\textbf{poet}) = \{a, c\} \neq \{a, b\} = \textbf{poet}$. And $h(\textbf{all}(\textbf{poet})) = h(\textbf{all})(h(\{a, b\})) = \textbf{all}(\{a, c\}) \neq \textbf{all}(\{a, b\}) = \textbf{all}(\textbf{poet})$, so the denotation of *all poets* is not fixed by all auts and so is not AI.

$\mathrm{AI}([P(E) \to \{0, 1\}])$ is small, of cardinality 2^{n+1} (Keenan and Moss 1985). We have some freedom since for $|E| = n$ and $\mathbf{d}$ an AI Det denotation, $(\textbf{d of the n})(E)$ is AI. In general the value of an AI function at an AI argument is AI: $h(f(\alpha)) = h(f)(h(\alpha)) = f(\alpha)$.

The set of restricting maps from $P(E)$ to $P(E)$ is exponential in $n = |E|$, and its AI subset is simply $1 + 2n$, linear in n (contrary to the figures given in Keenan and Moss 1984 and Keenan 1986).

Gen 2. Lexical items in small categories C denote freely in $\mathrm{Den}_E(C)$, those in large categories denote in a proper subset of $\mathrm{Den}_E(C)$.

The linguistically interesting part of Gen 2 concerns its second half. Consider first DPs, of which there are 2 to the 2^n possible denotations. Lexical DPs are primarily the Proper Nouns, which denote individuals and there are only $|E| = n$ such. In addition there are a small listable number of pronouns: *I, she*, etc. and a small listable number of quantified expressions (*everyone, someone, . . .*). So $|\mathrm{Lex}_E\mathrm{DP}| = n + c$ for c a (small) constant.

Lexical APs (not well understood) seem also to satisfy conditions stronger than restrictiveness which extensional APs in general satisfy. For example, if Al and Ed are both doctors and both lawyers, and Al is a tall doctor but Ed is not, then it cannot be that Ed is a tall lawyer but Al is not. But this condition may fail for complex APs such as *neither tall nor short*. Imagine a model with 10 doctors ordered uniformly by increasing height: $1 < 2 < \cdots < 10$. Let Al be 9, so he is a tall doctor, taller than 8 of the 10.

Let Ed be number 5, shorter then than half the doctors, so he is not a tall doctor. Suppose now that the lawyers are just the individuals $1, \ldots, 9$. Then Ed is a neither tall nor short lawyer, he is right at the midpoint. But Al is a tall lawyer, indeed the tallest, and so fails to be a neither tall nor short lawyer.

Finally consider Det_1s. They are usually increasing on their second argument: if $q \subseteq q'$ then $Dpq \leq Dpq'$. But there are some exceptions: *no* is decreasing on both its arguments. From KP/PK **no** is clearly lexical only in some Germanic languages (German *kein*; Danish *ingens*). More systematically some would treat simple numerals like *two* as denoting **exactly two**. So a more comprehensive but still substantive constraint on lexical Dets is Convexity (called Continuity in Thysse 1984):

Definition 4. D is *convex* iff whenever $q \subseteq q'$ and $Dpq = Dpq' = 1$ then for all s with $q \subseteq s \subseteq q'$, $Dps = 1$.

Dets such as *neither fewer than five nor more than ten* and *not between five and ten* are not convex.

A linguistic reflection   Being AI or ISOM is a mathematical property, a proper object of mathematical study, not a topic high on the ordinary linguist's list of concerns (but see below). It is then somewhat surprising that (in my opinion at least) we seem to have reliable intuitions concerning the distinction between expressions which are AI/ISOM and ones which are not. I'm not saying there are no unclear cases, but there are many clear ones, intuitively felt.

(2)   *Logical*   *Non-Logical*
    a.   *All* cats are grey   a'.   *John's* cats are grey
    b.   Ann *is* a surgeon   b'.   Ann *knows* a surgeon
    c.   Al cried *and* I fainted   c'.   Al cried *because* I fainted

So providing an even partial characterization of "logicality" in terms of AI/ISOM is a contribution of technical mathematics to our understanding of natural language.

Linguistic constants   As noted above, it is not an explicit goal of linguistic theory to characterize the notion *logical constant*. But linguistic theory does in practice distinguish *functional* expressions from *content* ones (which linguists call *functional categories* and *lexical categories* respectively). No formal definition of these notions is given, but the distinction is made in practice (see Emonds 1985 for some discussion).

Informally the linguist's functional expressions are ones that play a syntactic role in the grammar. E.g. the *-ly* that combines with adjectives to form adverbs: *smooth* ⇒ *smoothly, careful* ⇒ *carefully*, etc. is a functional expression. Much bound morphology, like *-ly*, is functional. But so are some single words: *the, a* and reflexive pronouns (*himself, herself*). Functional expressions are only replaceable by very few (often none) other expressions preserving grammaticality. Typological works, such as WALS, list languages that have a copula (*be*) or a definite article (*the*). No one classifies languages according as they have good translations of *forge*, or *recapitulate*.

Given our examples perhaps functional expressions in linguistics are those which denote AI objects. But that would let in too much, e.g. **most but not all**. Even limiting ourselves to AI lexical items seems too strong. **Most** does not seem to have any individuating syntactic function. So I suggest (Keenan and Stabler 2003):

Hypothesis The functional expressions in a language are those that are fixed by all automorphisms (structure maps) of the grammar.

There are two common ways linguists use to conceptualize a grammar G. The first: G is a finite lexicon (a set of categorized expressions) closed under a finite number of functions. The lexicon may be large, in the tens of thousands. The generating functions will be few. In current Minimalist theories we might have just Merge, or perhaps Internal Merge and External Merge. Perhaps Coordinate and Adjoin are additional functions. Then a structure map for L_G, the *language* generated by G, is just a bijection: $L_G \to L_G$ which fixes the generating functions. E.g. $h(\text{Merge}(x, y)) = \text{Merge}(h(x), h(y))$, etc.

This, it seems to me, is a pretty good characterization of what linguists actually mean by *functional*.

A more representational conception of a language is as a set of graphs, usually labeled trees, satisfying certain conditions. Then an automorphism of L is just a bijection from nodes to nodes which fixes the node labels and the Edge relation, i.e. there is an edge from n with label X to m with label Y iff there is an edge from $h(n)$ labeled X to $h(m)$ labeled Y. If the set of labels has structure (bar levels, slash depth) then we also require that this structure be preserved even if label identity is not required.

Now form the Type Hierarchy $\text{TH}(\text{Lex}_G, 2)$. An expression d is *syntactically invariant* (SI) iff for all (syntactic) structure maps h, $h(d) = d$. And properties like *having category C*, and relations, like *is a constituent of, is co-indexed with*, are SI iff they are fixed by all syntactic structure maps. And we can meaningfully ask:

(3) a. Do all syntactically invariant lexical items denote semantically invariant (AI) elements of their type?

 b. Is $PH(C)$, the set of phrases of category C, fixed by all syntactic structure maps? That is, must a structure preserving map map an expression of category C to one of category C? Does this hold for any (?all) C?

 c. Is there any regular relation between the syntactic invariants of one natural language and another?

These questions are addressed to varying degrees in Keenan and Stabler (2003). They give interestingly qualified negative answers to (3a,b). That some invariant lexical items also denote semantic constants seems reasonable (*not, the*) but others seem semantically rich and not "logically constant". For example causative affixes to verbs, illustrated below for Malagasy (Austronesian; Madagascar) seem to be a productive and highly integrated part of the grammar of many languages besides Malagasy (Chicewa, Turkish, Japanese):

(4) a. Sosotra   aho
         Irritated 1sg.nom

         I am irritated

 b. Mahasosotra   ahy   izany
         cause+irritated 1sg.acc that

         That makes me irritated

Regarding the invariance of grammatical categories, (3b) forces us to be aware of what kind of information is coded in subcategorization features. Superficially it would seem that an important role of grammatical categories is to assure similarity of syntactic distribution. But Keenan and Stabler illustrate a case where it would be formally possible to interchange, in a mini-model of Spanish, the masculine and feminine nouns, interchanging simultaneously masculine and feminine agreement markers, preserving grammaticality. The substitution seems a little "tricky" but it highlights the fact that gender marking has something symmetric about it, so it is not an accident of the grammar that such a systematic substitution could be possible.

Question (3c) is one that grammatical theory would like an affirmative answer to but, in my opinion, our current state of knowledge does not support this. We simply do not understand the grammar of any empirically given language well enough to make definitive statements about the structure of its set of syntactic invariants. In practice, generative grammarians tend to assume that different languages share the same gross grammatical architecture, though in practice we have no good way to compare the distributions of expressions in categories from different languages.

Chapter 11

Beyond the Frege Boundary

Chapter 4 concerned, in effect, type (2) functions built by composing two type (1) functions. So type (2) functions express properties of binary relations (and more generally map $k+2$-ary relations to k-ary ones). We see here that very few type (2) functions are compositions of type (1) functions (Keenan 1987c). We explore empirically other means English provides of expressing type (2) functions. Often in English, they are superficially similar to pairs of type (1) DPs, but on close inspection, identity fails.

We begin by considering expressions which can be adequately interpreted as type (2) functions H from $P_{E \times E} \to \{0, 1\}$, which are not equal to $F \circ G$ for any pair F, G of type (1) functions. Such H will be called *irreducible* (to type (1) functions; see Ben-Shalom 1994 for a more general statement and van Benthem 1989a for a much more extensive logical study). We exhibit an initial segment of a linguistic inventory, acknowledging that we have no systematic way of constructing such expressions. This is unsurprising, the expressions we seek are negatively defined — denoting functions that fail to have a certain property, and failure may have many causes.

Further, what is true syntactically is also true semantically. We will introduce a theorem which provides an applicable test for reducibility, but we do not know if there are any "positive" fully general properties that unreducible expressions satisfy. Some type (2) expressions can be analyzed as the value of a "Det" of type (1,1,2) which combines with two property denoting expressions to yield a type (2) denoting expression (jarringly discontinuous in English). These functions are appropriately conservative.

An informal inventory of type (2) expressions Reconsider first the iterated use of "standard" DPs and Dets that we studied earlier:

(1) a. No student answered every question (on the exam)

   b. $(\mathbf{no}(\mathbf{student}))((\mathbf{every}(\mathbf{question}))(\mathbf{answer}))$

    c.   (**no-every**)(**student, question, answer**)

    d. $(f - g)(p, q, R) = f(p)(g(q)(R))$

The semantic analysis in (1b) mirrors, roughly, the syntactic analysis in a natural way. That in (1c), interpreted as per (1d), seems inherently repugnant. No linguist would treat (*no, every*) as a constituent in (1a) so it should not be assigned a semantic interpretation at all. But (1c) is true in the same conditions as (1b), so the problem is only with what pieces of (1a) get assigned a meaning. And one advantage of the representation in (1c,d) is that it makes the effect of iterated conservativity easy to see:

(2)   Iterated Conservativity: $(f - g)(p, q, R) = (f - g)(p, q, R \cap (p \times q))$

Thus to check whether (1a) holds we do not have to check every pair in the **answer** relation but only those whose first coordinates are students and whose second are questions.

    Now consider (3), (3a) built on the pattern of the common expressions *Different people like different things / Different strokes for different folks.*

(3)   a.   *Different* students answered *different* questions (on the exam)

      b.   Ed danced with Ann but no one *else* danced with anyone *else*

      c.   *Which* students asked *which* questions (about the exam)?

The most natural reading of (3a) is one in which it imposes a weak one-to-oneness condition on the **answer** relation: "Pick any two students who took the exam, the set of questions one answered $\neq$ the set the other answered". Some might want to say that the two question sets are disjoint, but the weaker reading suffices for our later claims. *Different* also has a deictic or context dependent reading, as in *I interviewed a different student* (= different from one identified in context). We do not study such context dependent readings here or later. Sometimes *different* just seems to mean "diverse", or "at least two". The operator we study in (3a) is just the weak one-to-one one.

    In (3b) the *else-else* construction basically says that no pair besides (Ed,Ann) stands in the **dance with** relation. The second clause does not mean that only Ed danced with only Ann, which allows that people besides Ed danced with people other than Ann.

    In addition Moltmann (1992, 1996) notices that the *else-else* pattern applies to universally quantified expressions (4a), and that exceptives may induce type (2) quantifiers:

(4)   a.   Ed doesn't like Al but everyone else likes everyone else

    b.   Jeder Mann hat mit jeder Frau getanzt ausser Hans mit Maria
         Every man danced with every woman except Hans with Mary

(4a) means that only the pair (Ed, Al) is missing from the **like** relation. And (4b) states that only the pair (Hans, Maria) is absent from the **dance with** relation.

Lastly, the *which-which* question in (4c) asks us to identify the pairs of students and questions in the **ask** relation (with perhaps a preference for pairing with each student who asked a question all the questions s/he asked, but (4c) is not a paraphrase of *Which questions did each student ask?* which implies that each student asked at least one question, which (4c) does not.

And Nam (1996) notices that Gapping constructions in English induce type (2) quantifiers in both English and Korean. In the latter the two pairs of DPs are adjacent so their coordination seems natural.

(5)   a.   Every boy read a novel and every girl a play

    b.   Dan-i   sakwa-lul, kuliko Sue-ka   pay-lul   Bill-hantay cwuessta
         Dan-nom apple-acc and   Sue-nom pear-acc Bill-dat   gave
         Dan gave an apple and Sue a pear to Bill

Note now that the functions **Diff-Diff** and **Which-Which** of type (1,1,2) are conservative, where:

Definition 1. *H* of type (1,1,2) is *conservative* iff

$$H(p, q, R) = H(p, q, R \cap p \times q), \text{ all } p, q \subseteq E, \text{ all } R \subseteq E \times E.$$

The *diff-diff* construction surfaces in a variety of syntactic formats and extends to P_ns, $n > 2$ more generally:

(6)   a.   Mayors of different cities wore different colored neckties.

    b.   People from different countries have different tastes in food.

    c.   Different students answered different numbers of questions.

    d.   Different detectives interviewed different witnesses in different rooms.

Further the subject "antecedent" of an object-*different* need not itself be of the form (*a*) + *different* + Noun:

(7)   a.   Each pupil answered a different question (on the exam).

     b.   John and Bill attend different schools.

     c.   #Some student/#John answered a different question.

     d.   #Either Ed or Al/#Neither Ed nor Al attend different schools.

     e.   Ed and Tim have different accents/speak differently.

     f.   John and Bill respect each other.

(7a,b) suggest, wrongly, that *answer a different question / attend different schools* are ordinary P1s that can combine with type (1) DPs, such as *each pupil, John and Bill*. But (7c,d) show that type (1) DPs do not occur freely here. (The # indicates the absence of the one-to-oneness reading.) The examples suggest that $(a)+different+N$ is referentially dependent on the apparent subject, reminiscent of the reciprocal dependency in (7f). (7a) means that each pupil answered a question different from those answered by any of the other pupils.

And this leads us to observe that other Det/Adjectival expressions implying (non-)identity also induce unreducible type (2) functions:

(8)   a.   John and Bill support rival/opposing political parties.

     b.   They live in neighboring villages/the same village.

     c.   They live on opposite sides of the street/parallel streets.

     d.   They attend the same school and belong to the same clubs.

     e.   They live in the same country but ((live) in) different cities.

     f.   They live in the same country but not the same city.

     g.   They work in the same factory but don't live in the same city.

     h.   They work for the same bank but don't trust each other.

The later examples suggest boolean compounds of P1s of some type. Like the simpler cases, and like overtly reciprocal predicates, (8h), they require a plural *antecedent*. Replacing *they* by *he* above forces a context dependent reading or ungrammaticality. So interpreting these *same/different* P1s is not independent of the interpretation of the subject antecedent.

Note further that while *a different question* and *the same village* might seem to have the grammatical form Det+Adj+Noun, in fact it is better to think of *a different* and *the same* as syntactic units. The apparent Det cannot be freely replaced by other count Dets preserving a one-to-oneness reading:

(9) a. *John and Bill live in exactly one/each [same village].

 b. */#Max and Mack live in no/five [different villages].

Unsuspected ambiguities One merit of treating (**every, a different/the same**) etc as an operator of type (1,1,2) is that it provides a transparent way of expressing a variety of largely unnoticed ambiguities. Consider the interpretations of (10), writing R_a for $\{b \in E \mid (a,b) \in R\}$:

(10) a. Every pupil read the same three poems.

 b. (**every, same-3**)(**pupil, poem, read**)

 c. (**every, same$_3$**)$(p, q, R) = 1$ iff

   (i) for all $a, b \in p$, $q \cap R_a = q \cap R_b$ and
   (ii) $|q \cap R_a| = 3$

(10c;i) says that all pupils read the same poems, and (10c;ii) that the number of such poems was three. Other of our *same/different* type quantifiers can be given truth definitions in similar formats.

Consider now that (11) has four readings. They vary according to the relative scope of **Jon and Ben** and **two campers**, and as to which DP antecedes *same*. We exclude as always readings which are context dependent or group level. For example, the subject *Jon and Ben* antecedes *same* in *Jon and Ben told the same story to Ron*, but the indirect object antecedes it in *Sam told the same story to two campers*. (This is one point at which *same/different* may differ from overt reciprocals: **Ed described each other to two old friends*; but *??Ed described each other's families to two old friends who hadn't seen each other since childhood.*)

(11) Jon and Ben told the same story to two campers.

Here are four readings of (11). A formal statement using the format in (11) is given in an appendix. In R1 and R2, *two campers* outscopes *Jon and Ben*; in R3 and R4, *Jon and Ben* outscopes *two campers*.

R1:   There are two campers c, c' s.t. Jon told c the same story he told c' and so did Ben (the stories may be different)

R2:   There are two campers c, c' s.t. Jon told c the same story Ben did and Jon told c' the same story Ben did (stories may differ)

R3:   Jon told two campers the same story and so did Ben (both the campers and the stories may be different

R4:   Jon told two campers the same story as Ben told two campers (campers may be different, stories the same)

A test for reducibility (Keenan 1992)   Informally first, reducible functions are the same if they take the same values at the cross product relations.

Theorem 1. For all type (2) functions H and H', if both are reducible then $H = H'$ iff for all $P, Q \subseteq E$, $H(P \times Q) = H'(P \times Q)$.

Dekker (2003) generalizes this result to type (n) functions, ones that map n-ary relations to $\{0, 1\}$, and more generally $k + n$-ary ones to k-ary ones. Westerståhl (1994) generalizes still further and obtains a surprising result concerning iterations of type (1) that do not induce scope ambiguities. Ben-Shalom (1994) also generalizes to relations of arbitrary arity and presents a graph theoretical technique for representing various types of unreducibility.

To see how Theorem 1 works as a test, suppose we are to decide whether some type (2) H we have defined is reducible or not. We can show it unreducible if we can find type (1) functions F, G such that H and $F \circ G$ take the same value on the cross products but a different value on some other relation. Then one of H and $F \circ G$ is not reducible, and since $F \circ G$ trivially is we infer that H isn't.

As an example, let Refl be that function mapping binary relations R on E to 1 iff R is reflexive (that is, for all $x \in E$, $\langle x, x \rangle \in R$). Clearly $\text{Refl}(E \times E) = 1$, but that is the only product relation Refl is true of. If $P \times Q \neq E \times E$ then one of $P, Q \neq E$, so some x is missing so the pair $(x, x) \notin P \times Q$ so Refl maps it to 0. But $\mathbf{all}(E) \circ \mathbf{all}(E)$ maps $P \times Q$ to 1 iff $P = Q = E$. So Refl takes the same value at product relations as does $\mathbf{all}(E) \circ \mathbf{all}(E)$. But, with $|E| \geq 2$, these functions take different values at the identity relation $\{(x, x) \mid x \in E\}$. Thus Refl is unreducible since $\mathbf{all}(E) \circ \mathbf{all}(E)$ is reducible. Comparable arguments show that the maps sending a binary relation to 1 iff it is an equivalence relation, or a weak (or strict) partial order are unreducible. Consider now *different-different*:

(12)   Different pupils answered different questions (on the exam).

Choose E with at least 3 pupils: p_1, p_2, p_3 and two questions q_1, q_2. Let H be the type (2) (**diff pupil, diff question**) and consider its value on product relations $P \times Q$. If P lacks two or more of the pupils then there are two which, vacuously, answered the same questions (none), so $H(P \times Q) = 0$. If there are two students in P then they both stand in the answer relation to $Q \cap$**question** so again H maps $P \times Q$ to 0. This covers all the cases. Thus H takes the same values at the products as does the $\mathbf{0} \circ \mathbf{0}$ ($\mathbf{0}$ = **fewer than zero**, which maps all $P \subseteq E$ to 0). But $H(R) = 1$ when R is $\{(p_1, q_1), (p_2, q_2)\}$. So H is unreducible. Less obviously unreducible are *the same* cases:

(13) Ed and Ned answered the same two questions (on the exam).

The type (2) function H induced by the subject-object pair in (13) is, to the best of my judgment:

(14) $H(R) = 1$ iff

   a. Both Ed and Ned answered exactly two questions and

   b. The set of questions Ed answered = the set Ned answered

Now suppose $H(A \times B) = 1$. Then **ed** and **ned** are both in A, and B contains just two questions. By these conditions (14a) is also true, and if B contains either more than 2 or fewer than 2 questions the type (2) induced by (**Ed and Ned, exactly two questions**) is false of it, as is H. Thus H and (**Ed and Ned, exactly two questions**) take the same values on the product relations. But obviously they don't take the same value everywhere, as Ed and Ned might each have answered exactly two questions without them being the same. E.g. for **answer** $= \{(\mathbf{ed}, q_1), (\mathbf{ed}, q_2), (\mathbf{ned}, q_2), (\mathbf{ned}, q_3)\}$ where the q_j are distinct questions H is false but *Both Ed and Ned answered exactly two questions* is true. Thus H is unreducible. Several other induced H that we discuss below are similarly unreducible with just modest creativity involved in finding a reducible function that behaves the same on product relations.

Similarly, as Ben Shalom notes, even simple examples with indefinite antecedents are sometimes tricky to show unreducible:

(15) One boy insulted another boy (and then a fight broke out).

Suppose that $H = (\mathbf{one\ boy} - \mathbf{another\ boy})$ reduces to $F \circ G$. If **boy** includes $\{x, y\}$ with $x \neq y$ then H holds of $\{\langle x, y \rangle\}$, and of $\{\langle y, x \rangle\}$ as well as $\{\langle x, y \rangle, \langle y, x \rangle\}$ but not $\{\langle x, x \rangle, \langle y, y \rangle\}$ yet G maps these last two relations to the same set, hence $F \circ G$ maps them to the same truth value, which is false. Hence the putative reduction fails.

Proving unreducibility for expressions with *same* (or *similar*, as in *Darwin and Wallace presented similar theories in the same journal at the same time*) is sometimes subtle. More generally there are unreducible quantifiers that our test does not apply naturally to but Ben Shalom's graph techniques handle easily (including all those in Keenan 1992):

(16)   a.   Different teachers gave different tasks to different students.

       b.   Two guests introduced themselves to each other.

       c.   Ned gave his girlfriend more presents than Bill did.

Another observation that helps show unreducibility is:

Theorem 2. Red$_2$, the set of type (2) reducible H, is closed under complement, postcomplement and (thus) duals.

Proof. $\neg(f \circ g)(S) = \neg(f(g(S))) = (\neg f)(g(S)) = (\neg f \circ g)(S)$, whence $\neg(f \circ g) = (\neg f) \circ g$, so complements of reducible functions are reducible. The proof for postcomplements is similar. $\qquad\qquad\square$

Corollary. The unreducible H of type (2) are closed under complement, postcomplement and duals. (If H is unreducible and $\neg H$ is reducible then so is $\neg\neg H = H$, contradiction.)

The utility of Theorem 2 is seen in examples like (17) where H in (17a) is known to be unreducible and that in (17b) is its complement.

(17)   a.   Different students answered different questions.

       b.   No two students answered exactly the same questions.

The unreducible type (2) H are not closed under pointwise meets or joins. For H unreducible $H \wedge \neg H$ reduces to $\mathbf{0} \circ \mathbf{0}$ and $H \vee \neg H$ reduces to $\mathbf{1} \circ \mathbf{1}$. It is somewhat trickier to see that the reducible H are not closed under meets (or joins). Obviously the number of reducible H are upper bounded by the number of pairs $\langle f, g \rangle$ of type (1) functions. But not all pairs $\langle f, g \rangle$ induce different type (2) functions. One systematic collapse has already been discussed: the pairs $\langle f, g \rangle$ and $\langle f\neg, \neg g \rangle$ always induce the same function, so this cuts in half the upper bound above. In addition $\langle \mathbf{0}, f \rangle = \langle \mathbf{0}, g \rangle$ all type (1) f, g where $\mathbf{0}$ maps all p to 0. The actual figure (Keenan 1987c, p. 142) is:

Theorem 3. The number of reducible type (2) functions is $(\mathbf{a}-\mathbf{b})+4$, where $\mathbf{a}$ is 2 to the power $2^{n+1}-1$ and $\mathbf{b}$ is 2 to the power 2^n+1.

For example, with $|E|=2$, there are $2^{16}=65,536$ type (2) functions, just $2^7-2^5+4=128-32+4=100$ reducible ones.

Corollary. Red$_2$ is not closed under pointwise meets or joins.

Proof. Since Red$_2$ is closed under $\neg$, closure under $\wedge$ or $\vee$ implies closure under both, so Red$_2$ would be a finite boolean lattice (E finite) and thus complete and atomic, with cardinality 2^k for k the number of its atoms. But when $|E|=2$, $|\text{Red}_2|=100$, not a power of 2. So Red$_2$ is not closed under $\wedge$ or $\vee$. $\qquad\square$

Van Benthem (1989b, p. 462) provides a considerably stronger result in which we allow reduction not just by composition (iteration) as we have been doing but with full lambda abstraction.

The ability to detect unreducibility by checking product relations is surprising, as they constitute a very small portion of the binary relations. For $|E|=n$ there are $2^n \cdot 2^n = 2^{2n}$ products. But there are 2 to power n^2 binary relations. So for a 4 element universe there are $2^{16}=65,536$ binary relations only $2^8=256$ of which are products. So the latter are very special case. $R = P \times Q$ says that every P is related to every Q and nothing else is related to anything else. Defining product relations by lambda abstraction highlights their limited nature: R is a product iff $R = \lambda x(\lambda y(Px \,\&\, Qy))$. See also van Eijck (2005).

We turn now to two other types of expressions, *nominal and predicate anaphors*, which induce unreducible type (2) functions and have an independent role of interest in linguistic studies.

Nominal anaphors   Consider the interpretations of the object DPs in (18):

(18)   Each worker respects him / himself / his boss / his own boss.

The interpretation of the pronoun *him* is context dependent (hence not studied here), denoting some individual salient in context, like a free variable in first order logic. The interpretation of *himself*, an *anaphor* in linguistic parlance, is not context dependent. We interpret it here as **self**:

Definition 2. self$(R) = \{a \,|\, (a,a) \in R\}$

So *Each worker respects himself* is true iff **worker** $\subseteq$ **self**(**respect**). We note that **self** is AI (and ISOM) and that $(\mathbf{all}(E) \circ \mathbf{self})$ is not reducible as the only product relation it holds of is $E \times E$, like $\mathbf{all}(E) \circ \mathbf{all}(E)$. But it holds of the identity relation and the latter doesn't. $\mathbf{all}(E) \circ \mathbf{self}$ is also AI (ISOM) as the composition of AI (ISOM) functions is provably AI (ISOM).

Returning to (18), *his boss* is anaphorically non-committal. It might refer to the boss of someone salient in context, a context dependent interpretation (ignored here), or (18) might mean that for each worker w, w respects w's boss, in which case *his* and thus *his boss* is anaphoric. The use of *his own* tends to force an anaphoric interpretation, but sometimes strengthens it with a non-logical possession relation.

We have noted that many European languages present two possessives, one with a context dependent interpretation, a second with an anaphoric one: Latin *eius* vs *suus*; Norwegian *hans* vs *sin*; Russian *ego* vs *svoi*, Polish *jego* vs *swego* (Zuber 2010). See Keenan (2016). Also English bars bare anaphors — *himself, herself*, etc. from occurring as possessors: **John lost himself's wallet*, contrasting with *his, her*, etc. but many languages don't: Japanese, Mandarin, Korean, Hindi, Basque, ... And while English lacks these complex anaphors, it does present a variety of others (Keenan 1988; Safir 1992):

(19) a. Each worker criticized *every worker but himself.*

  b. No one likes to work with *anyone smarter than himself.*

  c. Each student criticized *both himself and the teacher.*

We want a way to say that the italicized expressions above are obligatorily interpreted as anaphoric, as is bare *himself* in their position. Doubtless their anaphoricity is due to the presence of *himself*, but that is not sufficient. The object DP in *I praised every student who criticized himself* is not interpreted anaphorically though it embeds an anaphor. So here we offer (Keenan 1986) an invariance condition that distinguishes among functions mapping binary to unary ($n+2$-ary to $n+1$-ary) relations according as they are anaphors or not.

Argument Anaphor Invariance   Let F map binary relations to unary ones. Then

  a. F is (*argument*) *free* (*a-free*) iff for all $a, b \in E$, all $R, S \subseteq E \times E$, if $R_a = S_b$ then $a \in F(R)$ iff $b \in F(S)$
    (This condition basically says that F is of type (1).)

    b. *F* is *argument bound (a-bound, anaphoric)* iff for all $a \in E$, all R, S
       if $R_a = S_a$ then $a \in F(R)$ iff $a \in F(S)$ and
       if in addition *F* is not free then *F* is a *proper anaphor*.   □

Most of Ted's teachers is a-free: if Ed admires just the individuals Rob fears then *Ed admires most of Ted's teachers* and *Rob fears most of Ted's teachers* must have the same truth value.

    In contrast **self** is not a-free. If Ed distrusts just Robin, Frieda, Ben, Zelda and Mick and those are just the individuals that Ben respects then *Ed distrusts himself* is false but *Ben respects himself* is true. But **self** is anaphoric: if Ed distrusts just the people he (Ed) respects then *Ed distrusts himself* and *Ed respects himself* have the same truth value. Similarly the complex object DPs in (19) denote proper anaphors. By contrast the function denoted by *every worker who criticized himself* is type (1) and thus a-free. So a DP with an embedded anaphor is not necessarily itself an anaphor.

    If *F* is anaphoric, whether *F* puts Ben in the set it associates with **respect** depends on <u>two</u> things: which individual Ben is (there are *n* choices), and which individuals he respects (2^n choices). If *F* is a-free *F* decides based solely on the second criterion. So our analysis enables us to show without stipulation that the complex objects in (19), *every worker but himself*, etc. are anaphors. Our claim that *himself (herself*, etc) in Ss like (18) denotes **self**, like our claim that *all* denotes **all,** etc.) is based on judgments of logical behavior, specifically entailment. E.g. the entailment ($\models$) patterns in (20) distinguish between a-bound *himself/herself* and a-free *him/her*, that is, between anaphors and pronominals in linguistic parlance.

(20)   a.   Some ball player hurt himself $\models$ Some ball player got hurt

      a′. Some ball player hurt him
           $\not\models$ Some ball player got hurt
           $\models$ Someone got hurt

      b.   Some student made herself quit smoking
           $\models$ Some student quit smoking

      b′. Some student made her quit smoking
           $\not\models$ Some student quit smoking
           $\models$ Someone quit smoking

An aside: our denotational approach to anaphora contrasts conceptually with standard Binding Theory (BT). BT just stipulates that lexical items like *himself, herself* and a listed handful of others are anaphors. This is of no help in deciding whether *caki* in Korean or *tena* in Malagasy are anaphors. On our approach expressions, lexical or not, are anaphors if they denote a-bound functions as defined above, like **self** or **no student but self**, which maps R to $\{b \in \textbf{student} \mid R_b \cap \textbf{student} = \{b\}\}$, is a-bound (anaphoric). Equally within standard BT there is no basis for arguing that expressions like those in (19) are anaphors (though the subconstituent *himself*, etc. is by stipulation). Whether expressions denote a-bound functions is decided by logical investigation, particularly entailment patterns, just as when we decide that *all* denotes **all**, etc.   *End aside*

Calling expressions *anaphors* if they denote a-bound functions as defined above, we observe that in English anaphors do not occur as subjects of main clause P_1s:

(21) *Himself / *No one but himself criticized every worker.

We might motivate the judgments in (21) on the grounds that both *himself* and *no one but himself* need an $n \geq 2$-ary relation as argument, but the syntax only makes a property, (**every worker**)(**criticized**), available. But all we can infer is that the subjects are not interpreted there as anaphors, not that the Ss are ungrammatical. The blocked pattern in (21) surfaces in Irish English (22); Jim McCloskey p.c.):

(22)   – Wait a minute! Herself is getting herself ready   (Irish)

The first *herself* in (22) is interpreted deictically as "prominent woman in context", the second anaphorically. Equally in Japanese (23b), Noriko Akatsuka p.c.) we find:

(23)   a.   Hanako ga   zibun   o   uttagatte-iru.
            Hanako nom self/Speaker acc doubt
            Hanako doubts herself / me

        b.   Zibun   ga   Hanako o   uttagatte-iru.
            Speaker/*self nom Hanako acc doubts
            I doubt Hanako
            *Hanako doubts herself

((23a,b) are in subordinate clause form, lacking the topic marker -*wa*).

Note too that many languages allow a designated anaphor to occur in the subject position of a complement clause bound to a higher subject: Japanese, Yoruba, Mandarin (24), examples from Wenyue Hua p.c.; see also Cole et al. (2017).

(24) a. Youxie ren   xiadao ziji le   / Youxie ren   xiadao ta   le.
          Some   people scared self past / Some   people scared him past

      b. Meige ren   dou juede ziji hen   congming.
          every people all   think self very smart
          Everyone$_i$ thinks he$_i$ is very smart

      c. Zhangsan$_i$ shi UCLA de xuesheng.
          Meige ren dou juede ta$_i$ hen congming.
          Zhangsan is a UCLA student
          Everyone thinks he's very smart

In fact in English from the 1300s into the 1700s, we find subject *himself/herself* either deictically interpreted (25c) or bound to a higher DP or one in a preceding clause (see Keenan 2002, 2009).

(25) (Dates for the first five are approximate.)

      a. There preached a Pardonere...   Piers Plowman 1370s
          And seide that *hymself* myght assoilen hem alle

      b. An he shal venge yow after that *hymself* witnesseth
                                                    Chaucer 1385

      c. Behold, *himself* has laid him down,   York Play 1425
          in length and breadth as he should be

      d. His knights grow riotous, and *himself* upbraids us
          On every trifle   Shakes. Lear I.3 1600

      e. Two glasses, where *herself herself* beheld
                                          Shakes. Venus and Adonis 1600

      f. ...a certain man ...giving out that *himself* was some great one
                                              King James 1611 Acts 8.9

      g. ...that stone which ...takes more room from others than *it selfe*
          fills
                                                    Hobbes 1651 1.78

      h. ...his Imperial Majesty, ...desired that *himself, and his Royal Consort* ...have the happiness ...of dining with me
                                                    Swift 1726 1.101

In sum: the anaphoric functions are closed under the pointwise boolean operations and each pair ⟨individual, property⟩ determines an atom so there are $n \cdot 2^n$ atoms, whence:

Theorem 4.

    a. $|\text{argument free}| = 2$ to the power 2^n

    b. $|\text{argument anaphor}| = 2$ to the power $n \cdot 2^n$

So in an E with just 3 elements there are $2^4 = 16$ a-free functions and $2^8 = 256$ anaphors (counting as degenerate cases the 16 free ones). So being able to denote anaphors increases the number of ways English can satisfy argument requirements of predicates. This is still a far cry from the full class of logical possibilities. But first we note a linguistic advantage of the semantic approach taken here:

Gen 1. All natural languages can express argument anaphors.

    What this means in our context is that all Ls have lexical expressions (affixes, clitics, full words) that can be interpreted as **self**. But not all have expressions which must be so interpreted. Old and Middle English (Keenan 2002, 2009) did not. OE used *hine* as an accusative masc. sg. pronoun that could be interpreted as **self**, (26), but also allowed antecedents in the previous discourse, (27):

(26)   ...þæt   he   moste   mid   feo   hine   alysan   Bede c890
           ...that   he   must   with   money   himself   ransom

(27)   forðæm   nan mon_i ne   bitt oderne_j ðæt he_j hine_i
           because   no man_i asks another_j that he_j him_i lift

                rære,   gif he_i self   nat   ðæt he_i afeallen   bið
                up   if he_i (self)   not-knows   that he_i fallen   be

                                                      Cura Pastoralis c800

Dative *him* replaced *hine* in later Middle English but took both local and non-local antecedents, as did the newly created *-self* forms.

    But recall, for $|E| = 2$, there are $2^{32} > 4$ billion maps from binary to unary relations, so our language still suffers a lack of expressive power. One way we can express more is with *predicate-anaphors* (p-anaphors) illustrated in (28):

Predicate anaphors (P-anaphors)

(28)   a.   Leslie read *more novels than Pete.*

         b.   Leslie interviewed *the same candidates as Pete.*

(28a) means that Leslie read more novels than Pete read. Replacing *read* with *burned* the result means that Leslie burned more novels than Pete burned. So the object in (28a), and (28b) varies with the denotation of the main P_2. These expressions satisfy:

(29)   PREDICATE INVARIANCE: if $R_a = R_b$ then $a \in F(R)$ iff $b \in F(R)$

So if Leslie read the same objects that Bill read then *Leslie read more novels than Pete* iff *Bill read more novels than Pete*. Ditto if she and Bill interviewed the same individuals.

So called *antecedent contained deletion* is another type of p-anaphor:

(30)   Leslie read every novel that Ruth did.

Here we might think of *did* as a P_2 variable bound by *read*. It cannot be grammatically deleted in (30), whereas its presence is optional in (28a,b). And clearly if Leslie and Lolly read exactly the same things then Leslie read every novel that Ruth did iff Lolly read every novel that Ruth did, so *every novel that Ruth did* is a p-anaphor.

Like a-anaphors, p-anaphors do not occur in English as the sole argument of a P_1:

(31)   a. *Every student who Ned does knows Pete
         (Intended: Every student who Ned knows knows Pete)

       b. *More students than Ned does know Pete
          (Intended: More students than Ned knows know Pete)

       c. *The same students as Jeb does know Pete
          (Intended: The same students as Jeb knows know Pete)

Theorem 5. $|\text{Predicate Anaphor}| = |[P(E \times E) \to \{0,1\}]| = 2$ to the power $2^{n \cdot n}$.

So for $|E| = 2$ there are $2^{16} = 65,536$ p-anaphors, massively more than the $2^8 = 256$ a-anaphors, but still massively less than the 2^{32} maps from binary to unary relations.

Effability again   Is English expressive enough to denote (over finite E) any function from binary relations to properties? I think so. The objects in (32) can vary both with the relation and its subject argument, which should give us enough power to denote the atoms in $[P(E \times E) \to P(E)]$ but we have not worked this out explicitly.

(32) a. $Mary_i$ read more poems than / the same poems as her_i mother.

  b. $Ruth_i$ interviewed every candidate that her_i friend did.

In addition the authors we have cited discuss additional types of type (2) functions. See Ben-Shalom (1994), Peters and Westerståhl (2006), and Szymanik (2010) for both extensive review and new observations. Here we just note *cumulative* quantifiers, as in (33) and *resumptive* quantifiers (type (2,2) for example), as in (34).

(33) Three assistants graded more than 100 exams (between them).

(34) Most neighbors are friends. Most twins never separate.

Szymanik (2010) shows that cumulative and resumptive quantifiers stay within the bounds of polynomial complexity whereas branching and Ramsey quantifiers (not studied here) exceed this bound.

Conclusion Early generative grammar used rules like S → NP + VP, and VP → V + NP. These rules, while not false, are misleading, as, without further additions they invite the inference that objects of transitive verbs are the same as subjects of VPs. Semantically however this is a colossal misrepresentation. The range of denotable functions from binary to unary relations massively exceeds that of the functions from unary relations to truth values e.g. a-anaphors, p-anaphors and many other unreducible type (2) DPs.

Appendix

Here are the formal representations of the four readings. I find them hard to read on first pass. We write L2 for **at least two**.

R1: $L2(\mathbf{camper}, \{c \,|\, \text{Jon \& Ben told the same story to } c\})$
    $= L2(\mathbf{camper}, \{c \,|\, (\mathbf{every}, \mathbf{same}_1) \, (\{\mathbf{j}, \mathbf{b}\}, \mathbf{story}, \{(x, y) \,|\, x \text{ told } y \text{ to } c\})\})$

R2: $(L2, \mathbf{same}_1)(\mathbf{camper}, \mathbf{story}, \{(x, y) \,|\, \textbf{John \& Ben told } y \textbf{ to } x\})$

R3: $\mathbf{every}(\{j, b\}, \{a \,|\, a \text{ told the same story to two campers}\})$
    $= \mathbf{every}(\{j, b\}, \{a \,|\, (L2, \mathbf{same}_1)(\mathbf{camper}, \mathbf{story}, \{(b, c) \,|\, a \text{ told } b \text{ to } c\})\})$

R4: $(\mathbf{every}, \mathbf{same}_1)(\{j, b\}, \mathbf{story}, \{(a, b) \,|\, a \text{ told } b \text{ to two campers}\})$

Chapter 12

Eliminating the Universe

This chapter is adumbrated in Keenan (1982) and builds on Keenan (2015), from which the semantic problem we address is taken.

Our concern here is to model the validity of a certain productive entailment pattern that forces a distinction between extensional and intensional interpretations of common noun phrases (CNPs) and their modifiers. To this end we offer a novel notion of model, of which standard extensional models are a special case. We compare our approach with a "possible worlds" approach to intensional interpretation. We note that the intensional phenomena which arise in our case are "model internal" and do not, at least obviously, invite comparisons with other states of affairs, in contrast to verbs of propositional attitudes (*believe*) and non-extensional verbs (*seek*).

Our presentation is in three parts: (1) the empirical entailment paradigm of interest; (2) summary of possible worlds models of the paradigm; (3) the boolean models we propose. They are built solely on known boolean structure and do not take as primitive either a set E of entities or a set W of possible worlds (or any other unknowns, such as propositions).

The entailment paradigm   We consider *value judgment* modifiers of *agentive* CNPs. The latter denote *properties* of individuals who regularly engage in an activity in which one can be proficient to varying degrees. This is a highly productive class in English, exemplified in (1).

(1)   a.   surgeon, poet, lawyer, doctor, author, midwife

b.   novelist, flautist, logician, mathematician, philosopher, linguist

c.   heart surgeon, opera singer, pool player, race car driver, mountain climber, portrait painter, civil rights lawyer

Such CNPs may be (synchronically) monomorphemic as in (1a), or bimorphemic built from a root+suffix such as *-ist*, *-er* or *-ian*. They may also incorporate an object as in (1c). In addition, and the focus of our interest here, such CNPs may be modified, unboundedly, by adjectival phrases expressing a value judgment concerning how well participants perform the relevant activity. Such modifiers may be pre- or post-nominal. We sometimes accompany the CNP with an indefinite article *a/an* to improve readability.

(2)   A *skillful* heart surgeon, a *gifted* mathematician, a *talented* flautist

(3)   A *very skillful and conscientious* heart surgeon, a *very talented but often inattentive and careless* flautist, an *exceptionally gifted* psychiatrist *wrongly thought to be indifferent to his patients*

Restrictive modifiers. We represent value judgment modifiers as *restricting* functions F from properties to properties, meaning that the individuals with $F(p)$ are a subset, not necessarily proper, of those with the Noun property p. Thus the a-sentences below entail their a′-counterparts, so these are semantic facts an analysis of these CNPs and modifiers must account for.

(4)   a.   Rob is a skillful surgeon $\models$ a′.   Rob is a surgeon

    b.   Dana is a talented flautist $\models$ b′.   Dana is a flautist

    c.   Sue is a first rate physicist $\models$ c′.   Sue is a physicist

Non-extensionality. Value judgment modifiers are non-extensional: the heart surgeons and the portrait painters may be the same individuals in some context (model) but it does **not** follow that the skillful heart surgeons and the skillful portrait painters are the same in that context (model). Thus it may happen in some model that {Dana, Robin, Luke, Sue, Sean} is the set of heart surgeons as well as the set of portrait painters, but the set of skillful heart surgeons is {Dana, Robin, Luke} and the set of skillful portrait painters is {Sue, Robin}. We shall refer to the set of individuals with a property p in a model M as the *extension* of p in M, noted $\mathrm{ext}_M(p)$, often omitting M when not comparing models. The entailment paradigm we account for here, using F as an arbitrary value judgment function, is:

(5)   a. F is restricting:
        $\mathrm{ext}_M(F(p)) \subseteq \mathrm{ext}_M(p)$, all models M.

    b. F is non-extensional:
        for some M, $\mathrm{ext}_M(p) = \mathrm{ext}_M(q)$ but $\mathrm{ext}_M(F(p)) \neq \mathrm{ext}_M(F(q))$.

Worth noting, though not pursued here, is that the non-extensionality judgment in (5b) extends beyond basic adjectival modification:

(6) The opera singers and the poker players may coincide but

    a. The best opera singer and the best poker player may differ.

    b. The top two opera singers and the top two poker players may differ.

Standard models + possible worlds As we have seen, standard models of semantic relations in simple English build on two semantic primitives: truth and reference. So a model for our assumed language is a triple: $(E, \{0,1\}, [\![\cdot]\!])$, E any non-empty set of entities, $\{0,1\}$ the boolean lattice of truth values with $1 =$ **True** and $0 =$ **False**, and $[\![\cdot]\!]$ a function mapping lexical expressions (non-logical constants) into elements of $\mathrm{TH}(E, \{0,1\})$, the type hierarchy generated by E and $\{0,1\}$. We often use boldface $\mathbf{d}$ for $[\![d]\!]$. To understand the need to extend this notion of model to handle the non-extensional phenomena of concern we look more closely at the very different semantic primitives E and $\{0,1\}$. E, the *universe* of a model, is any set, taken to be non-empty as a minor convenience. E might be finite or (un)countably infinite. It has no structure — it is not endowed with any functions, relations or designated elements. So different models may have universes of different cardinalities and models with the same universe may be non-isomorphic.

In contrast $\{0,1\}$ is structured, supporting various functions and relations. It supports a partial order relation $\leq$: for all $x, y \in \{0,1\}$, $x \leq y$ iff $x = 0$ or both x and $y = 1$. As noted earlier, this is just the material implication relation. It fails only when $x = 1$ and $y = 0$. Also we might say that 1 is the value a conjunction of sentences has iff each conjunct has that value, and it is the value a disjunction of sentences has iff at least one of the disjuncts has that value. This perspective enables us to see that the actual identity of elements of the set in which sentences denote is immaterial. Any two element set, say $\{\mathrm{F}, \mathrm{T}\}$ or $\{\beta, \alpha\}$, would do, as long as one of the elements was stipulated to have the properties we attributed to 1 above. But there would be little gain in allowing the elements of $\{0,1\}$ to vary across models as any two finite boolean lattices of the same cardinality are isomorphic and thus make the same sentences true.

To see that this notion of model is not rich enough to capture the entailment paradigm in (5) consider that if the same individuals have the property denoted by *heart surgeon* and that denoted by *poker player* then the properties denoted by these CNPs must be the same, since uti (up to isomorphism) a property is just a set of individuals (an element of $P(E)$, equivalently, $[E \to \{0,1\}]$, the set of functions from E into $\{0,1\}$). Thus if value judgment adjectives like *skillful* denote functions from properties to properties they must yield the same value at **heart surgeon** and **poker player** in this case, violating the entailment paradigm in (5).

So semanticists enrich the notion of a model M to include a new parameter W_M, subscript often omitted, whose elements are (poetically) called "possible worlds". Then we interpret a sentence φ in a model M, noted $[\![\varphi]\!]^M$, as a function from W_M into $\{0,1\}$. We refer to this function as the *intension* of φ and its value at any $w \in W_M$ as its *extension* in w. So the extensions of sentences are truth values. Of course the function $[\![\cdot]\!]^M$ is designed to capture standard constraints on boolean connectives and quantifiers. E.g. $[\![\varphi \,\&\, \psi]\!]^M(w) = [\![\varphi]\!]^M(w) \wedge [\![\psi]\!]^M(w)$, and $[\![\forall x\varphi]\!]^M(w) = \bigwedge\{[\![\varphi]\!]^M(w') \mid w' \in W_M\}$.

With this apparatus we can straightforwardly capture the entailment paradigm in (5). Interpret CNPs as functions from W into $P(E)$, called *intensional properties*. So the extension of *heart surgeon* in a world $w \in W$ is just the set of entities in E that $[\![heart\ surgeon]\!]^M$ assigns to w. And now we interpret adjectival phrases such as (*very*) *skillful* as restricting functions from intensional properties to intensional properties which are not required to be extensional (transparent), where these notions are defined below:

Definition 1. For f a function $[W \to P(E)]$ into $[W \to P(E)]$

    a. f is *restricting* iff for all $p \in [W \to P(E)]$, all $w \in W$, $(f(p))(w) \subseteq p(w)$

    b f is *extensional* iff for all $p, q \in [W \to P(E)]$, all $w \in W$ if $p(w) = q(w)$ then $(f(p))(w) = (f(q))(w)$

One checks that for E and W with at least two elements each there are restricting non-extensional maps as per the definition above. We propose a different solution to representing the entailment paradigm in (5), so let us first reflect on the merits and costs of the possible worlds approach just sketched.

Two important merits of Possible Worlds Semantics (PWS)

    1. The PWS above captures, for the data range considered, the idea that the intension of an expression determines its extension.

    2. The PWS above does capture the entailment paradigm in (5).

Four qualms re Possible Worlds Semantics

My qualms here concern solely my attempt to use PWS to capture the entailment paradigm in (5), specifically concerned with evaluative adjectives applied to agent nominals. PWS has much broader applications in modal logic and the analysis of propositional attitudes, none of which are under consideration here. And very possibly my attempt to use PWS to represent evaluative adjectives is not what others would have chosen. That said, here are my qualms:

1. PWS invokes a new unknown: W, a set of "possible worlds". This is, in my opinion, a last resort strategy:

**To understand something new characterize it
in terms of notions that are already understood.**

As a last resort we may on occasion have to posit a novel force to explain something. But if we did that regularly we would have as many forces as falling apples. In general we count as advances in understanding cases in which phenomena originally treated as disparate come to be understood as special cases of more general phenomena. The laws of fluid mechanics subsume those of hydrodynamics and aerodynamics; electric fields and magnetic fields are now part of a larger electromagnetic spectrum, etc.

It is of course quite possible that the intensional phenomena we discuss do require a new unknown. But as a matter of good science we should try first to characterize them in terms of known conceptual or mathematical notions. As a rule in theory construction we support:

SCIENTIFIC IMPERATIVE
Minimize Unknowns

Our analysis of the entailment paradigm in (5) satisfies this Imperative.

2. The fact that the truth of sentences varies with possible worlds is confusing as the truth of sentences also varies with the models. Now the latter claim is quite reasonable as our semantics is compositional and the interpretation of *John* and *smokes* can vary from model to model (even with the same universe) so we find unsurprisingly that *John smokes* is true in some models and false in others. But why should the truth of that sentence vary with some random index? It is unsatisfying to "explain" one mystery by invoking another.

This variation also clouds our representation of entailment. Classically we say that φ entails ψ iff for all models M, if $[\![\varphi]\!]^M = 1$ then $[\![\psi]\!]^M = 1$. The idea is that no matter what the world is like, if it is the way φ says it is then it is the way ψ says it is. But with PWS the intuition is that φ entails ψ iff for all models M and all $w \in W_M$ if $[\![\varphi]\!]^M(w) = 1$ then $[\![\psi]\!]^M(w) = 1$ (Montague 1970, p. 208). This statement is less clear than the original as we have no way to assess whether *John smokes* (or the axiom of choice, or. . .) is true in some w_6 or not. So PWS is, in my judgment, too committed to inscrutable notions.

3. The appropriate notion of the intension of a CNP should be more than just a list of its possible extensions. Let us think informally, not inconsistent with PWS, of the intension of CNP denotations p, as what you must learn to know that an arbitrary individual has p. So the intension of **heart surgeon** is the criterion that must be satisfied for someone to be a heart surgeon. This notion of intension is cognitive, knowledge based, and thus is at least the kind of thing that can be expressed in language. And it supports a natural inclusion relation:

Definition 2 (informal). For p, q CNP intensions, $p \leq q$ iff knowing that an arbitrary individual x has p implies knowing that x has q.

An easy example: the intension of **skillful and accomplished surgeon** $\leq$ the intension of **skillful surgeon**. If you have figured out that Fred is a skillful and accomplished surgeon then you have figured out that he is a skillful surgeon.

Using the intuition expressed by the definition it is easy to see that $\leq$ is reflexive: knowing that an individual has p implies knowing that that individual has p. Equally transitivity is straightforward, and antisymmetry is reasonable: if knowing that x has p implies knowing that x has q and vice versa then to know one is to know the other, so they are not cognitively distinct: $p = q$. Thus the $\leq$ relation is a partial order: reflexive, transitive, and antisymmetric.

And if knowing that an x has s implies knowing that x has p and also that x has q then, trivially, it implies knowing that x has both p and q, that is, $p \wedge q$. And again plausibly knowing that an x has p and also q or has r implies knowing that x has p and q or that it has p and r, so the relation behaves distributively. Complements are a little different from what we have seen so far. To decide that x is a not very skillful surgeon we must know that x is a surgeon who fails to be a very skillful one (not simply anything that fails to be a very skillful surgeon, like my cat for example). Lastly, consistent with these observations is the existence of a tautologous property $\mathbf{1}$, one where we don't really have to know anything to know that an arbitrary individual has that property. So vacuously all CNP intensions $p \leq \mathbf{1}$.

These considerations support that the inclusion relation among intensions has a boolean character — that of a distributive complemented lattice. This much could also be expressed in a PWS, as we show here, but the boolean structure turns out, wrongly, to be atomic, just "lifted" from the extensional structure of $P(E)$:

Theorem 1. For $(B, \leq)$ any boolean lattice (BL) and A a non-empty set, $([A \to B], \leq)$ is a BL, where $\leq$ is given by: For all $f, g \in [A \to B]$, the set of functions from A into B, $f \leq g$ iff for all $a \in A$, $f(a) \leq g(a)$.

Corollary. If $(B, \leq)$ is atomic so is $([A \to B], \leq)$

(For each $x \in A$, f_x is an atom in $([A \to B], \leq)$, where $f_x(y) = 1_B$ iff $y = x$. And for each $g \neq 0$ (0 maps all x to 0_B) there is an $f_x \leq g$.)

So a PWS which captures the boolean properties of CNPs and their modifiers assigns a complete atomic boolean structure to the set of CNP intensions, fully analogous to $P(E)$. I claim this is the wrong structure. Since $P(E)$ is atomic so is $[W_M \to P(E)]$. But the natural CNP intension relation is not atomic. It differs from other instances of boolean relations we have seen in that we can add without bound modifiers to a non-zero CNP obtaining ones with increasingly strong conditions an x has to meet to have the property: *a surgeon, a heart surgeon, a skillful heart surgeon, a very skillful heart surgeon, a very skillful and compassionate heart surgeon, a very skillful and compassionate but not very accomplished heart surgeon,* etc. The modifiers may grow in length and heaviness but there seems to be no natural cut-off point. We may continually restrict without getting to a property that cannot be further restricted. Boolean lattices (algebras) with this character are well known and easily comprehensible as they are built around meaning relations in languages. Their elements can be considered as *information packets*. The most basic one, and all we need here, is the one built from the formulas of SL (Sentential Logic).

Recall that SL is built from denumerably many basic formulas $P_1, P_2, \ldots$ under appropriate combinations with *and, or, not,* etc. A model for SL is a function m mapping Basic Formulas to $\{0, 1\}$, the minimal BL (boolean lattice). For each such m, an *interpretation* of SL is a function m^* mapping all the formulas in SL into $\{0, 1\}$ in such a way that its values on the basic formulas coincide with those m assigns them and their values on syntactically complex formulas are given recursively by the truth tables for *and, not,* etc. E.g.

$$m^*(\varphi \,\&\, \psi) = m^*(\varphi) \wedge m^*(\psi), \, m^*(\text{not } \varphi) = \neg m^*(\varphi), \text{ etc.}$$

For $\varphi, \psi \in \text{SL}$, write $\varphi \equiv \psi$ to say that φ is *logically equivalent* to ψ, that is, for all models m, $m^*(\varphi) = m^*(\psi)$. Write $[\varphi]$ for $\{\psi \mid \varphi \equiv \psi\}$. $[\varphi]$ is an *information packet* — the set of formulas that have the same logical meaning as φ, that is, are true in the same models as φ. Note: $[\varphi]$ is always infinite since for τ any tautology (a formula like *(P1 or not P1)* which all m^* map to 1) φ, $(\varphi \,\&\, \tau)$, $((\varphi \,\&\, \tau) \,\&\, \tau)$, ... are logically equivalent (There are less trivial examples).

Now let us write LT for $\{[\varphi] \mid \varphi \in \text{SL}\}$, the set of information packets over SL. (LT stands for Lindenbaum-Tarski.) Then,

Theorem 2. $(LT, \leq)$ is a boolean lattice, where $[\varphi] \leq_{\text{def}} [\psi]$ iff for all models m, $m^*(\varphi) = 1 \to m^*(\psi) = 1$. (See Bell and Slomson 1971, pp. 40–42.)

We note that $[\varphi] \wedge [\psi] = [(\varphi \,\&\, \psi)]$, $[\varphi] \vee [\psi] = [(\varphi \text{ or } \psi)]$, and $\neg[\varphi] = [\text{not } \varphi]$. The zero element is the equivalence class of a contradiction and the unit the class of a tautology. (These are well definitions as $\equiv$ is a congruence relation, per Bell and Slomson above.)

Of interest is the theorem below. It tells us that for any φ in SL which is true in some model there is an infinite descending sequence $[\varphi], [\varphi'], [\varphi''], \ldots$ such that each later packet is strictly more informative than all the prior packets, that is, each later packet $\leq$ all the prior packets and no prior packet $\leq$ it. This means that we can move down the sequence eliminating some models in which the formulas in the previous set were true without ever getting to the empty set of models.

To see how this is true let φ be any formula with a model. Let k be the largest integer such that the basic formula P_k occurs in φ. (There is such a k since φ contains only finitely many symbols.) Then $[\varphi] > [(\varphi \,\&\, P_{k+1})] > [((\varphi \,\&\, P_{k+1}) \,\&\, P_{k+2})], \ldots$ ($[\varphi] > [\psi]$ means $[\psi] < [\varphi]$, which abbreviates $[\psi] \leq [\varphi]$ and $[\psi] \neq [\varphi]$). Thus the set of models that make ψ true is a proper subset of that which make φ true. So as we move down the sequence we become increasingly informative still leaving ourselves uncommitted to the truth of infinitely many basic formulas. It is this unbounded increase in informativity that makes LT algebras appropriate models of value judgment modifiers. For the formal record:

Definition 3. An atom α in a BL $(B, \leq)$ is a non-zero element with the property that for all β in B, $\beta \leq \alpha \Rightarrow \beta = 0$ or $\beta = \alpha$. Thus an atom in a boolean lattice is a smallest non-zero element.

Theorem 3. $(LT, \leq)$ is atomless.

Proof. Let $[\varphi]$ be a non-zero element of LT. Then φ has a model m, and there is a greatest k such that the basic P_k occurs in φ. Define m' to be that model which maps P_{k+1} to 0 and all other P_n to whatever m maps them to. Then $m'^*(\varphi) = m^*(\varphi) = 1$ but $m'^*(\varphi \,\&\, P_{k+1}) = 0$, so $[\varphi \,\&\, P_{k+1}] < [\varphi]$. And m'' which maps P_{k+1} to 1 and all other P_n to $m(P_n)$ makes $(\varphi \,\&\, P_{k+1})$ true, so $0 < [\varphi \,\&\, P_{k+1}]$. Thus all non-zero elements of LT are not atoms, so LT is atomless. $\qquad\square$

Corollary. Every atomless boolean lattice is infinite.

Proof. Each non-zero element heads an infinite descending chain as illustrated above. $\qquad\square$

Finally each language for which truth in a model is defined the logical equivalence classes form a boolean LT lattice.

4. On the PWS sketched (drawn on Montague 1970) the intensions associated with a given CNP vary with the model, as the set of possible worlds varies from model to model. But on the intuition of intension presented here this seems not to be the case. What you have to know to know that an x is a skillful surgeon is unaffected by fluctuations in the surgeon population. It could be that the last skillful surgeon passed away last night, all the rest are just hacks. But to justify our claim that there are no longer any skillful surgeons we still need to know what properties someone must have to be one.

Boolean models: no possible worlds, no entities (!) Here we propose models with just two semantic primitives: the BL $\{0, 1\}$ of truth values and a BL P of *properties*, denotations for CNPs, common noun phrases. So all our primitives are uniformly boolean. Proper noun denotations, which we call *individuals*, are definable in terms of P and are justified in terms of judgments of entailment on simple English sentences.

Our presentation comes in two parts. First we show that we can formulate standard extensional models with just the two primitives P and $\{0, 1\}$. Our models make the same sentences true as standard ones taking E and $\{0, 1\}$ as primitive. So the elimination of the universe as a primitive notion does not in and of itself have any radical consequences. It delightfully satisfies our dictum:

If you can't say something two ways, you can't say it

And the reformulated standard models do have some consequences of potential philosophical (Keenan 2015) and cognitive relevance.

Second, we show that we can model the intensional phenomena in the entailment paradigm in (5) just by weakening one of the conditions the primitive P is required to satisfy. These conditions are instantiated by well known boolean algebras, so they satisfy our criterion of not adding new unknowns to represent phenomena we don't understand.

Old models in new garb   Classical extensional models, as we have seen, build on the boolean set $\{0, 1\}$ of truth values and an arbitrary (non-empty) set E of entities. E provides denotations for proper nouns (individual constants in mathematical languages) and is the range of individual variables (for languages which have them). Users of these languages speak of proper nouns (individual constants) as *referring* to entities in E. In contrast CNPs, such as *surgeon, skillful surgeon*, etc. denote sets of entities, elements of $P(E)$, the power set of E. Since $P(E)$ is isomorphic to $[E \rightarrow \{0, 1\}]$ — the function mapping each $K \subseteq E$ to the function f_K which maps any b in E to 1 iff $b \in K$ is an isomorphism — we can equally well treat CNPs as denoting in $[E \rightarrow \{0, 1\}]$ without changing which sentences are shown to entail which others.

Now $(P(E), \subseteq)$ is a specific sort of boolean lattice, a complete atomic one. An atom in $P(E)$ is a smallest non-zero element, the zero element being the empty set $\varnothing$. In general the zero element of a BL is that unique element $\leq$ all the others. So for each $b \in E$, $\{b\}$ is an atom of $P(E)$, and $P(E)$ is atomic, meaning that for each non-zero (non-empty) element X there is an atom $\{b\} \subseteq X$. (Obviously if X is non-empty then there is a $b \in X$ so $\{b\} \subseteq X$.) If we take the set of CNP denotations to be $[E \rightarrow \{0, 1\}]$ an atom is a function which maps exactly one b in E to 1 and all the others to 0.

A BL $(B, \leq)$ is *complete* iff each subset K of B has a glb (greatest lower bound) and also a lub (least upper bound). Recall that x is a lb (lower bound) for K iff for all $y \in K$, $x \leq y$, and in addition x is greatest of the lbs for K iff for all lbs z for K, $z \leq x$. Dually u is an ub (upper bound) for K iff for all $v \in K$, $v \leq u$, and in addition u is a lub for K iff for all ubs z for K, $u \leq z$. For all $x, y \in B$, $\{x, y\}$ has a glb, usually noted $x \wedge y$, and a lub, usually noted $x \vee y$. By induction any finite subset of a BL has a glb and a lub, but infinite subsets may fail to have glbs and lubs. If a subset K of B has a glb (lub) it is noted $\bigwedge K$ ($\bigvee K$). In $P(E)$, given any collection A of subsets of E, $\bigwedge A$ is just the intersection of the sets in A, that is, the set of elements of E that lie in each set in A. And $\bigvee A$ is the union of the sets in A, that is, the collection of b which lie in at least one of the sets in A. And we note:

Theorem 4. A BL $(B, \leq)$ is complete and atomic iff for some set E, $(B, \leq)$ $\simeq (P(E), \subseteq)$.

Proof. We have already seen that $P(E)$ is ca (complete and atomic), and if $(B, \leq)$ is ca then it is isomorphic to $(P(\text{Atom}(B), \subseteq)$, where $\text{Atom}(B)$ is the set of atoms of B. The map sending each x in B to $\{\alpha \in \text{Atom}(B) \,|\, \alpha \leq x\}$ is an isomorphism. □

Isomorphism between structures plays an important role in what follows. Our motivation is the following informally given theorem:

Theorem 5. Isomorphic mathematical structures make the same sentences true (Enderton 1972, p. 92)

Here the mathematical structures of concern are boolean lattices (algebras) and the sentences are from the mathematical language of boolean lattices, in which we can say things like $\neg(x \wedge y) = (\neg x \vee \neg y)$ and $x \leq y$ iff $x \wedge \neg y = 0$, etc.

Proper nouns and the universe E   Given Theorem 5, anything in the type hierarchy generated by E and $\{0, 1\}$ that we could build from $\{0, 1\}$ and $P(E)$, we can build, up to isomorphism, starting with $\{0, 1\}$ and an arbitrary ca boolean lattice P. What seems to be missing on the latter approach is E itself. For natural language purposes, its primary function is to provide denotations for proper nouns (*Dana*, etc.).

Now proper nouns are a special case of DPs, which also include expressions like *every student, Ruth but no teacher*, etc. In general they denote functions (generalized quantifiers) from properties, P, into $\{0, 1\}$, which we do represent in our new models. And Determiners, such as *every, no*, ... are functions from properties to these functions, elements of $[P \rightarrow [P \rightarrow \{0, 1\}]]$, also definable from just P and $\{0, 1\}$.

And it turns out that proper nouns can be discriminated from other DPs by properties of the functions that they can denote. Consider first the logical equivalences in (7).

(7)   a.   Dan isn't vegan $\equiv$ It is not so that Dan is vegan

   b.   Al either laughed or cried $\equiv$ Either Al laughed or Al cried

   c.   Sam is brave and is loyal $\equiv$ Sam is brave and Sam is loyal

Now treating the proper nouns as functions taking the predicates as arguments observe the elegant boolean pattern in (8):

(8)   a.   Dan(not vegan) ≡ not(Dan(vegan))

     b.   Al(laugh or cry) ≡ (Al(laugh) or Al(cry))

     c.   Sam(wise and loyal) ≡ (Sam(wise) and Sam(loyal))

In boolean terms (8a) says that a proper noun maps the complement of a P_1, a one place predicate, to the complement of the truth value you get when you apply the proper noun function to the (unnegated) predicate. Similarly in (8b) the proper noun maps a disjunction (lub) of predicates to the lub of the truth values obtained by applying the proper noun function to each of the predicates. And dually in (8c) proper nouns map a glb of predicates to glbs of the truth values.

(9)   Proper nouns denote complete homomorphisms in $[P \rightarrow \{0,1\}]$.

So proper nouns form a proper subset of DPs and denote in a proper subset of the functions from P into $\{0,1\}$.

Even within the classical notion of model we can easily treat proper nouns as generalized quantifiers and prove they are homomorphisms. (Other DPs sometimes denote homomorphisms: in a model in which there is just one poet, *every poet*, *some poet*, and *the poet* all denote that individual. Singular definite descriptions such as *the person who discovered penicillin* denote individuals or the zero.)

Definition 4. For each $b \in E$ define $I_b \in [P(E) \rightarrow \{0,1\}]$ by: $I_b(p) = 1$ iff $b \in p$. Equivalently $I_{\{b\}}(p) = 1$ iff $\{b\} \subseteq p$, given that $b \in p$ iff $\{b\} \subseteq p$.

Then we verify that these I_bs commute with the boolean operations:

(10)   a. $I_d(\neg\mathbf{vegan}) = \neg(I_d(\mathbf{vegan}))$

     b. $I_a(\mathbf{laugh} \vee \mathbf{cry}) = I_a(\mathbf{laugh}) \vee I_a(\mathbf{cry})$

     c. $I_s(\mathbf{wise} \wedge \mathbf{loyal}) = I_s(\mathbf{wise}) \wedge I_s(\mathbf{loyal})$

To see (10a), note: $I_d(\neg\mathbf{vegan}) = 1$ iff $d \in E - \mathbf{vegan}$, iff $d \notin \mathbf{vegan}$), iff $I_d(\mathbf{vegan}) = 0$, iff $\neg(I_d(\mathbf{vegan}) = 1$. (10b,c) follow similarly and we see that individuals map BLs to BLs in such a way as to preserve the boolean operations. That is, proper nouns denote homomorphisms. (A homomorphism maps a BL to a BL in such a way that $h(x \wedge y) = h(x) \wedge h(y)$; $h(x \vee y) = h(x) \vee h(y)$ and $h(\neg x) = \neg(h(x))$.) Other DPs we have considered do not in general denote homomorphisms. *Each baby either laughed or cried* does not have the same logical meaning as *either each baby laughed* *or each baby cried; Exactly one student here isn't vegan $\not\equiv$ It is not so that exactly one student here is vegan.*

It turns out that completely respecting the boolean operations suffices to characterize individuals. A homomorphism h here is *complete* if its value at an intersection of a bunch of subsets of E is the glb of the set of its values at the sets in the bunch. In more detail, I_b is complete iff it preserves arbitrary glbs and lubs. That is, if p_j is a CNP property for each j in some set J then $I_b(\bigcap\{p_j \mid j \in J\}) = \bigwedge\{I_b(p_j) \mid j \in J\}$ (=1 iff for all $j \in J$, $I_b(p_j) = 1$). And dually, $I_b(\bigcup\{p_j \mid j \in J\}) = \bigvee\{I_b(p_j) \mid j \in J\}$ (=1 iff for some $j \in J$, $I_b(p_j) = 1$).[1]

Theorem 6 (Keenan and Westerståhl 1997, fn. 3). A function h from $P(E)$ into $\{0,1\}$ is a complete homomorphism iff for some $b \in E$, $h = I_b$.

Interpreting proper nouns as elements $b \in E$ or as I_bs does not change truth conditions for simple sentences: On the former account $[\![John\ smokes]\!]$ = **smoke**($\mathbf{j}$) on the latter it = $I_j(\mathbf{smoke})$ =$_{\text{def}}$ **smoke**($\mathbf{j}$), already clear from work in Montague grammar. To conclude:

(11)   Standard models with primitives $(E, \{0,1\})$, E an arbitrary (non-empty) set, and uniform ones with primitives $(P, \{0,1\})$, P a ca BL are elementarily equivalent — they support the same judgments of entailment between sentences.

A restatement of Theorem 6 using an arbitrary ca boolean lattice P is:

Theorem 7. A function h from a ca P into $\{0,1\}$ is a complete homomorphism (c-hom) iff for some atom $\alpha \in P$, $h = I_\alpha$, where $I_\alpha(p) = 1$ iff$_{\text{def}}$ $\alpha \leq p$.

Note that for any cardinal $n > 0$ there is a ca boolean lattice P with exactly n atoms ($P(E)$, for $|E| = n$) and hence n individuals, so taking a ca P as primitive does not limit the number of individuals, and eliminating the universe does not induce any radical changes. Why then engage in this algebraic high jinks? The general answer is that a new way of formulating something can lead to new questions, new observations, and new generalizations. And this we show here. Lastly the reader may wonder how we provide denotations for P_ks in general without a universe. The easy answer is just to use

---

[1] Requiring that P be complete is a convenience, not a necessity. There is a canonical way of minimally extending any boolean lattice to a complete one (Givant and Halmos 2009, p. 216, also exercise 2 p. 219) which, crucially does not add any new atoms.

We define a homomorphism from $(B, \leq)$ into $(D, \leq)$ to be *complete* iff it preserves whatever glbs and lubs exist. That is, if $K \subseteq B$ has a glb then $\{h(k) \mid k \in K\}$ is a subset of D with a glb in D and $h(\bigwedge_B K) = \bigwedge_D \{h(k) \mid k \in K\}$; dually for joins.

Atom(P) in place of E. So whereas earlier P_{k+1}s denoted maps from E into P_k denotations recursively now they just denote maps from Atom(P) into P_k denotations. That is, for $k > 0$, Den$_P(P_{k+1}) = [\text{Atom}(P) \to \text{Den}_P(P_k), \leq]$, provably a ca BL pointwise.

Ultimately though we would like an abstract characterization of n-ary relations analogous to the way in which the set of unary relations (properties) is characterized as a complete atomic BL. McKinsey (1940) provides this for binary relations but I know of no characterization for the set of n-ary relations in general. See Jonsson (1991) for an introduction to binary relation algebras.

And even characterizing the unary relations abstractly we find a split between the relational lattice theoretic and the algebraic perspective. An algebraic operation on the domain of a BL can define an n-ary glb operation for any finite $n > 0$, but we cannot algebraically define an operator that takes arbitrary subsets of the domain to their glb. That invokes proper (monadic) Second Order Logic.

Generalizing models $(P, \{0, 1\})$ to intensional ones   To generalize $(P, \{0, 1\})$ to represent the intensional phenomena in (5) we do not have to add anything, such as possible worlds, or propositions. It suffices to *weaken* the requirements on P, specifically the atomicity requirement. If P is not required to be atomic we will have infinite descending sequences of properties which can model unbounded addition of value judgment modifiers without reaching the contradictory property, as in the LT lattices. But by our previous discussion we also want some atoms so that various DPs can denote individuals (complete homomorphisms).

So we would like the boolean lattice P in which CNPs denote to have at least one atom but to be non-atomic. Do such boolean lattices exist? Yes, as follows: For $(B, \leq_B)$ and $(D, \leq_D)$ any boolean lattices, their cross product $(B \times D, \leq)$ is standardly a BL; the $\leq$ relation is defined in parallel: $(b, d) \leq (b', d')$ iff $b \leq_B b'$ & $d \leq_D d'$. It follows that the boolean operations are also run in parallel: e.g. $(b, d) \wedge (b', d') = (b \wedge_B b', d \wedge_D d')$, and $\neg(b, d) = (\neg_B b, \neg_D d)$. The zero element is $(0_B, 0_D)$ and the unit $(1_B, 1_D)$. (b, d) is an atom of $B \times D$ iff b is an atom of B and d the zero of D, or b the zero of B and d an atom of D.

Now consider the product of $P(E)$ with LT (of Sentential Logic). This lattice has atoms: $(\{b\}, 0_{\text{LT}})$, $b \in E$. Thus we can interpret proper nouns and other DPs as individuals satisfying entailment paradigms like that in (8) and (10).

But the cross product lattice is not atomic: for d a non-zero element of LT, $(0_B, d)$ is a non-zero element of the cross product lattice but there is no atom $(\{b\}, 0_{\text{LT}}) \leq (0_B, d)$. So the product lattice is not atomic. We define:

Intensional models are triples $M = (P, \{0, 1\}, [\![\cdot]\!])$, where P is a complete[2] non-atomic boolean lattice with at least one atom, $\{0, 1\}$ is the truth value lattice, and $[\![\cdot]\!]$ is a function from expressions into the elements of $\text{TH}(P, \{0, 1\})$ discussed below.

For $p \in P$, we define $\text{ext}(p)$, the *extension* of p, as $p \wedge \bigvee \text{Atom}(P)$, and $\text{int}(p)$, the *intension* of p, as $p \wedge \neg \bigvee \text{Atom}(P)$. $0 < \bigvee \text{Atom}(P)$ since there is at least one atom in P. Also $\bigvee \text{Atom}(P) < 1_P$ since by a general theorem for any BL B (complete or not), B is atomic iff 1_B is the lub of $\text{Atom}(B)$. Thus $\neg \bigvee \text{Atom}(P)$ also lies strictly between 0_P and 1_P. This implies that $P \simeq \,\downarrow\! \bigvee \text{Atom}(P) \times \,\downarrow\! \neg \bigvee \text{Atom}(P)$, where for any BL B and any $x \in B$, $\downarrow\! x = \{y \mid y \leq x\}$. And for $0 < x < 1$, $\downarrow\! x$ is a BL with unit x. The $\leq_{\downarrow x}$ is just $\leq_B$ restricted to $\downarrow\! x$.[3]

$\downarrow\! \bigvee \text{Atom}(P)$ is ca, and $\downarrow\! \neg \bigvee \text{Atom}(P)$ is complete and atomless. So each CNP denotation p can be thought of as a pair $(\text{ext}(p), \text{int}(p))$, where $\text{ext}(p)$ is its atomic part, $\text{ext}(p) \leq \bigvee \text{Atom}(P)$, and $\text{int}(p)$ is its atomless part, $\text{int}(p) \leq \neg \bigvee \text{Atom}(P)$. It is immediate that for $p, q \in P$, $p \leq q$ iff $\text{ext}(p) \leq \text{ext}(q)$ and $\text{int}(p) \leq \text{int}(q)$. It is thus clear that even for p with a very small extension, p can head an infinite descending sequence as long as $\text{int}(p) \neq 0_{\downarrow \neg \bigvee \text{Atom}(P)}$.

The interpreting function $[\![\cdot]\!]$ is required to satisfy two conditions:

(12) a. For d a value judgment modifier, $[\![d]\!]$ is restricting:
          For all CNP denotations p, $\text{ext}([\![d]\!](p)) \leq \text{ext}(p)$, and

  b. For all CNP denotations p, q,
          if $\text{int}(p) \leq \text{int}(q)$ then $\text{ext}(p) \leq \text{ext}(q)$

(12a) has been motivated. (12b) derives from our pretheoretical definition of the implication relation $\leq$ between CNP denotations: if knowing that an x is a p implies knowing that it is a q then an x that is a p ought to also be a q, otherwise we couldn't know that it is a p. (So we are accepting some sort of general knowability assumption.)

(12b) allows that $p < q$, that is, $p \leq q$ & $p \neq q$, even if $\text{ext}(p) = \text{ext}(q)$. And this is reasonable. It might be that we have a model in which all heart surgeons are skillful ones, so $\text{ext}(\textbf{heart surgeon}) = \text{ext}(\textbf{skillful heart}$

---

[2]See footnote 1 on page 149.
[3]So $\wedge$ and $\vee$ and 0 in $\downarrow\! x$ coincide with what they are in B. The unit in $\downarrow\! x$ is x, and $\neg_{\downarrow x} y = x \wedge \neg_B y$.

surgeon) even though **skillful heart surgeon** < **heart surgeon**, the reason being that int(**skillful heart surgeon**) < int(**heart surgeon**).

Conclusion + a last model theoretic constraint   Our analysis does treat value judgment modifiers as restricting and non-extensional, and, in terms of (12b) it uses intensions to constrain extensions, though less strongly than listing them "world by world".

Re our qualms concerning PWS our analysis satisfies the first three directly: There is no new unknown, W; truth of sentences varies only with models, and we admit of arbitrarily long restrictions on properties without forcing a corresponding extensional restriction. But we have not yet addressed qualm 4, that intensions may vary with the model. Surely we want extensions to so vary, as different (numbers of) individuals may exist in different states of affairs. But the criteria for deciding that an x is a heart surgeon seem not to vary with models. They are part of the fixed language we are modeling, more like truth values than individuals.

So for the moment let us require that in any model $M = (P, \{0, 1\}, [\![\cdot]\!])$, P is up to isomorphism $P_{ext} \times P_{int}$, where P_{ext} and P_{int} are both complete boolean lattices, P_{ext} atomic and P_{int} atomless.

Bibliography

Arnauld, Antoine and Claude Lancelot (1975). *General and Rational Grammar: The Port-Royal Grammar.* Original 1660, translated J. Rieux and B. E. Rollin. The Hague: Mouton.

Bach, Emmon, Eloise Jelinek, Angelika Kratzer, and Barbara Partee, eds. (1995). *Quantification in Natural Language.* Dordrecht: Kluwer.

Barwise, Jon (1979). "On branching quantifiers in English." In: *Journal of Philosophical Logic* 8, pp. 47–80.

Barwise, Jon and Robin Cooper (1981). "Generalized quantifiers and natural language." In: *Linguistics and Philosophy* 4, pp. 159–219.

Beghelli, Filippo (1994). "Structured quantifiers." In: *Dynamics, Polarity and Quantification.* Ed. by Makoto Kanazawa and Christopher Piñón. CSLI Lecture Notes, pp. 119–145.

Bell, John L. and Alan B. Slomson (1971). *Models and Ultraproducts.* Amsterdam: North Holland.

Ben-Shalom, Dorit (1994). "A tree characterization of quantifier reducibility." In: *Dynamics, Polarity and Quantification.* Ed. by Makoto Kanazawa and Christopher Piñón. CSLI Lecture Notes, pp. 173–212.

Boole, George (1854). *An Investigation of the Laws of Thought (On which are founded the Mathematical Theories of Logic and Probabilities).* Reprint: the Open Court Pub. Co., LaSalle, Illinois.

Boolos, George (1981). "For every A there is a B." In: *Linguistic Inquiry* 12, pp. 465–466.

Boolos, George, John Burgess, and Richard Jeffrey (2007). *Computability and Logic.* Cambridge University Press.

Boolos, George and Richard Jeffrey (1980). *Computability and Logic*. 2nd edition. Cambridge: Cambridge University Press.

Burnett, Heather (2014). "A delineation solution to the puzzles of absolute adjectives." In: *Linguistics and Philosophy* 37.1, pp. 1–39.

Chomsky, Noam (1956). "Three models for the description of language." In: *IRE Transactions on Information Theory* 2.3, pp. 113–124.

— (1957). *Syntactic Structures*. Mouton.

— (1959). "On certain formal properties of grammars." In: *Information and Control* 2.2, pp. 137–167.

— (1975). *Reflections on Language*. Pantheon Books.

— (1986). *Knowledge of Language*. Praeger.

— (1966/2009). *Cartesian Linguistics*. Cambridge University Press.

Chomsky, Noam and Howard Lasnik (1977). "Filters and control." In: *Linguistic Inquiry* 8, pp. 425–504.

Cole, Peter, Gabriella Hermon, and C. T. James Huang (2017). "Long Distance Reflexives." Ms.

De Saussure, Ferdinand (1916). *Cours de linguistique générale*. Payot.

De Swart, Henriëtte (1996). "Quantification Over Time." In: *Quantifiers, Logic, and Language*. Ed. by Jaap van der Does and Jan van Eijck. Stanford: CSLI, pp. 311–337.

Dekker, Paul (2003). "Meanwhile, within the Frege boundary." In: *Linguistics and Philosophy* 26.5, pp. 547–556.

Ebbinghaus, Heinz-Dieter, Jörg Flum, and Wolfgang Thomas (1984). *Mathematical Logic*. Springer.

Emonds, Joseph (1985). *A Unified Theory of Syntactic Categories*. Dordrecht: Foris.

Enderton, Herbert B. (1972). *A Mathematical Introduction to Logic*. New York: Academic Press.

Epstein, Richard and Walter Carnielli (1989). *Computability*. Wadsworth & Brooks/Cole.

Etchemendy, John (1990). *The Concept of Logical Consequence*. Cambridge, MA: Harvard University Press.

Fauconnier, Gilles (1975a). "Pragmatic scales and logical structure." In: *Linguistic Inquiry* 6.3, pp. 353–375.

— (1975b). "Do quantifiers branch?" In: *Linguistic Inquiry* 6, pp. 555–578.

Frege, Gottlob (1893). *Grundgesetze der Arithmetik 1*. Partial translation and introduction by M. Furth, *The Basic Laws of Arithmetic*. University of California Press. Berkeley, 1964.

Gabbay, Dov and Julius Moravcsik (1974). "Branching quantifiers, English, and Montague grammar." In: *Theoretical Linguistics* 1, pp. 141–157.

Givant, Steven and Paul Halmos (2009). *Introduction to Boolean Algebras*. Springer.

Graf, Thomas, Denis Paperno, Anna Szabolcsi, and Jos Tellings, eds. (2012). *Theories of Everything*. Vol. 17. UCLA Working Papers in Linguistics. UCLA.

Grätzer, George (1998). *General Lattice Theory*. Birkhäuser.

[WALS] Haspelmath, Martin, Matthew Dryer, David Gil, and Bernard Comrie, eds. (2005). *World Atlas of Language Structures*. Oxford: Oxford University Press.

Heim, Irene and Angelika Kratzer (1998). *Semantics in Generative Grammar*. Oxford: Blackwell.

Hintikka, Jaako (1973). "Quantifiers vs. quantification theory." In: *Linguistic Inquiry* 5, pp. 153–177.

Jackendoff, Ray (1972). *Semantic Interpretation in Generative Grammar*. MIT Press.

— (1990). *Semantic Structures*. Cambridge, MA: MIT Press.

Jesperson, Otto (1924). *The Philosophy of Grammar*. London: George Allen & Unwin.

Johnsen, Lars G. (1987). "There sentences and generalized quantifiers." In: *Generalized Quantifiers*. Ed. by Peter Gärdenfors. Springer, pp. 93–109.

Jonsson, Bjarni (1991). "The theory of binary relations." In: *Algebraic Logic*. Ed. by Hajnal Andreka, J. Donald Monk, and István Németi. North-Holland, pp. 245–293.

Keenan, Edward L. (1982). "Eliminating the universe (A study in ontological perfection)." In: *Proceedings of the First West Coast Conference on Formal Linguistics*. Ed. by Daniel Flickinger, Marlys Macken, and Nancy Wiegand. Stanford Linguistics Association, pp. 71–81.

— (1986). "Lexical freedom and large categories." In: *Studies in Discourse Representation Theory and the Theory of Generalized Quantifiers*. Ed. by Jeroen Groenendijk, D. de Jongh, and Martin Stokhof. Dordrecht: Foris Publications, pp. 27–53.

— (1987a). "A semantic definition of indefinite NP." In: *The Representation of (In)Definiteness*. Ed. by E. Reuland and A. ter Meulen. Cambridge, MA: MIT Press, pp. 286–317.

— (1987b). "Multiply-headed NPs." In: *Linguistic Inquiry* 18.3, pp. 481–490.

— (1987c). "Unreducible *n*-ary quantifiers in natural language." In: *Generalized Quantifiers*. Ed. by Peter Gärdenfors. Springer, pp. 109–150.

— (1988). "Complex Anaphors and Bind Alpha." In: *Papers from the 24th Annual Regional Meeting of the Chicago Linguistic Society*. Ed. by I. Macleod, G. Larson, and D. Brentani. Chicago: CLS, pp. 216–232.

— (1992). "Beyond the Frege Boundary." In: *Linguistics and Philosophy* 15, pp. 199–221.

— (1993). "Natural Language, Sortal Reducibility and Generalized Quantifiers." In: *Journal of Symbolic Logic* 58.1, pp. 314–325.

— (1996). "The semantics of determiners." In: *The Handbook of Contemporary Semantic Theory*. Ed. by S. Lappin. Oxford: Blackwell, pp. 41–63.

— (2000). "Quantification in English is inherently sortal." In: *History and Philosophy of Logic* 20, pp. 251–265.

— (2001). "Logical Objects." In: *Logic, Meaning and Computation*. Ed. by C. Anthony Anderson and M. Zelëny. Springer, pp. 149–180.

— (2002). "Explaining the creation of reflexive pronouns in English." In: *Studies in the History of the English Language*. Ed. by Donka Minkova and Robert Stockwell. Mouton de Gruyter, pp. 325–354.

— (2005). "Excursions in natural logic." In: *Studies in Mathematical Linguistics and Natural Language*. Ed. by Claudia Casadio, Philip Scott, and Robert Seely. CSLI, pp. 3–24.

— (2008). "Further excursions in natural logic: the midpoint theorems." In: *Logics for Linguistic Structures*. Ed. by Fritz Hamm and Stephan Kepser. Mouton de Gruyter, pp. 87–104.

— (2009). "Linguistic theory and the historical creation of English reflexives." In: *Historical Syntax and Linguistic Theory*. Ed. by P. Crisma and G. Longobardi. Oxford: Oxford University Press, pp. 17–41.

— (2015). "Individuals explained away." In: *On Reference*. Ed. by Andrea Bianchi. Oxford: Oxford University Press, pp. 384–401.

— (2016). "*In Situ* Interpretation without Type Mismatches." In: *Journal of Semantics* 33.1, pp. 87–106.

Keenan, Edward L. and Leonard Faltz (1985). *Boolean Semantics for Natural Language*. Dordrecht: Reidel.

Keenan, Edward L. and Lawrence S. Moss (1984). "Generalized Quantifiers and the Logical Expressive Power of Natural Language." In: *WCCFL 3*. Ed. by Mark Cobler, Susannah MacKaye, and Michael T. Westcoat. Stanford: Stanford Linguistic Association, pp. 149–157.

— (1985). "Generalized Quantifiers and the Expressive Power of Natural Language." In: *Generalized Quantifiers in Natural Language*. Ed. by Johan van Benthem and Alice ter Meulen. Dordrecht: Foris Publications, pp. 73–127.

— (2016). *Mathematical Structures in Language*. CSLI.

[KP] Keenan, Edward L. and Denis Paperno, eds. (2012). *Handbook of Quantifiers in Natural Language, vol. 1*. Springer.

Keenan, Edward L. and Edward P. Stabler (2003). *Bare Grammar*. Stanford: CSLI.

Keenan, Edward L. and Jonathan Stavi (1986). "A semantic characterization of natural language determiners." In: *Linguistics and Philosophy* 9, pp. 253–326.

Keenan, Edward L. and Dag Westerståhl (1997). "Generalized Quantifiers in Linguistics and Logic." In: *Handbook of Logic and Language*. Ed. by Johan van Benthem and Alice ter Meulen. North Holland, pp. 859–910.

Klima, Edward (1964). "Negation in English." In: *The Structure of Language*. Ed. by J. Fodor and J. Katz. Englewood Cliffs, NJ: Prentice-Hall, pp. 245–323.

Ladusaw, William (1983). "Logical form and conditions on grammaticality." In: *Linguistics and Philosophy* 6, pp. 389–422.

Lappin, Shalom (1988). "The semantics of 'many' as a weak quantifier." In: *Linguistics* 26, pp. 977–998.

Lewis, David (1970). "General semantics." In: *Synthese* 22, pp. 18–67.

Lindström, Per (1966). "First order predicate logics with generalized quantifiers." In: *Theoria* 32.3, pp. 186–195.

— (1969). "On extensions of elementary logic." In: *Theoria* 35.1, pp. 1–11.

Linebarger, Marcia (1987). "Negative polarity and grammatical representation." In: *Linguistics and Philosophy* 10, pp. 325–387.

Liu, Feng Hsi (1996). "Branching quantification and scope independence." In: *Quantifiers, Logic, and Language*. Ed. by Jaap van der Does and Jan van Eijck. Stanford: CSLI, pp. 155–168.

Löbner, Sebastian (1986). "Quantification as a major module of natural language semantics." In: *Studies in Discourse Representation Theory and the Theory of Generalized Quantifiers*. Ed. by Jeroen Groenendijk, D. de Jongh, and Martin Stokhof. Dordrecht: Foris Publications, pp. 53–87.

Loveland, Donald W., Richard E. Hodel, and S. G. Sterrett (2014). *Three Views of Logic: Mathematics, Philosophy, and Computer Science*. Princeton University Press.

Matthewson, Lisa (2008). *Quantification: a Cross-Linguistic Persepctive*. Bingley.

McKinsey, J. C. C. (1940). "Postulates for the calculus of binary relations." In: *Journal of Symbolic Logic* 1.3, pp. 85–97.

McNally, Louise (2011). "Existential sentences." In: *Semantics: An International Handbook*. Ed. by C. Maienborn, K. von Heusinger, and P. Portner, pp. 1829–1848.

— (2016). "Existential sentences cross-linguistically: variations in form and meaning." In: *Annual Review of Linguistics* 2, pp. 211–231.

Milsark, Gary (1977). "Toward an explanation of certain peculiarities in the existential construction in English." In: *Linguistic Analysis* 3, pp. 1–30.

Moltmann, Friedericke (1992). "Coordination and Comparatives." PhD thesis. MIT.

—   (1996). "Resumptive quantification in exception sentences." In: *Quantifiers, Deduction, and Context*. Ed. by Makoto Kanazawa, Chris Piñón, and Henriëtte de Swart. Stanford: CSLI, pp. 139–170.

Montague, Richard (1970). "English as a Formal Language." In: *Formal Philosophy: Selected Papers of Richard Montague*. Ed. by R. Thomason. Originally presented at the 1970 Stanford Workshop on Grammar and Semantics. New Haven, CT: Yale University Press, pp. 247–271.

—   (1973). "The proper treatment of quantification in ordinary English." In: *Formal Philosophy: Selected Papers of Richard Montague*. Ed. by R. Thomason. New Haven, CT: Yale University Press, pp. 188–221.

Nam, Seungho (1996). "N-ary quantifiers and the expressive power of NP compositions." In: *Quantifiers, Logic, and Language*. Ed. by Jaap van der Does and Jan van Eijck. Stanford: CSLI.

[PK]   Paperno, Denis and Edward L. Keenan, eds. (2017). *Handbook of Quantifiers in Natural Language. Vol 2*. Springer.

Pederson, Eric (1993). "Zero negation in South Dravidian." In: *CLS 27, papers from the 27th regional meeting of the Chicago Linguistics Society 1991, part two: Parasession on Negation*. Ed. by L. Dobrin, L. Nichols, and R.M. Rodriguez. Chicago: CLS, pp. 233–245.

Peters, Stanley and Dag Westerståhl (2006). *Quantifiers in Language and Logic*. Oxford: Oxford University Press.

Putnam, Hilary (1981). *Reason, Turth and History*. Cambridge: Cambridge University Press.

Rando, Emily and Donna Jo Napoli (1978). "Definites in *there*-sentences." In: *Language* 54.2, pp. 300–313.

Rotstein, Carmen and Yoad Winter (2004). "Total vs. partial adjectives: scale structure and higher-order modifiers." In: *Natural Language Semantics* 12, pp. 259–288.

Safir, Ken (1992). "Non-coreference and the pattern of anaphora." In: *Linguistics and Philosophy* 15.1, pp. 1–52.

Suihkonen, Pirkko (2007). *On Quantification in Finnish*. Lincom GmbH.

Suihkonen, Pirkko and Valery Solovyev, eds. (2013). *Typology of Quantification*. Lincom GmbH.

Szabolcsi, Anna and Frans Zwarts (1997). "Weak islands and an algebraic semantics for scope taking." In: *Ways of scope taking*. Ed. by Anna Szabolcsi. Kluwer Academic Publishers, pp. 217–263.

Szymanik, Jakub (2010). "Computational complexity of polyadic lifts of generalized quantifiers in natural language." In: *Linguistics and Philosophy* 33, pp. 215–250.

Tarski, Alfred (1931). "The concept of truth in formalized languages." In: *Logic, Semantics and Metamathematics*. Ed. by J. H. Woodger. Clarendon Press, pp. 152–278.

Ter Meulen, Alice (1995). *Representing Time in Natural Language*. Cambridge: MIT Press.

Ter Meulen, Alice and Eric Reuland, eds. (1987). *The Representation of (In)Definiteness*. MIT Press.

Thysse, Elias (1984). "Counting quantifiers." In: *Generalized Quantifiers in Natural Language*. Ed. by Alice ter Meulen and Johan van Benthem. Dordrecht: Foris, pp. 127–147.

Van Benthem, Johan (1986). *Essays in Logical Semantics*. Dordrecht: D. Reidel.

— (1989a). "Logical constants across varying types." In: *Notre Dame Journal of Formal Logic* 30.3, pp. 315–342.

— (1989b). "Polyadic quantifiers." In: *Linguistics and Philosophy* 12.4, pp. 437–465.

— (2010). *Modal Logic for Open Minds*. CSLI.

Van Eijck, Jan (2005). "Normal forms for characteristic functions." In: *Journal of Logic and Computation* 15.2, pp. 85–98.

Vendler, Zeno (1968). *Adjectives and Nominalizations*. The Hague: Mouton.

Westerståhl, Dag (1985). "Logical constants in quantifier languages." In: *Linguistics and Philosophy* 8, pp. 387–413.

— (1989). "Quantifiers in formal and natural languages." In: *The Handbook of Philosophical Logic*. Ed. by D. Gabbay and F. Guenthner. Vol. IV. Dordrecht: Reidel, pp. 1–131.

— (1994). "Iterated quantifiers." In: *Dynamics, Polarity and Quantification*. Ed. by Makoto Kanazawa and Christopher Piñón. CSLI Lecture Notes, pp. 147–171.

— (2012). "Midpoints." In: *Theories of Everything*. Ed. by Thomas Graf, Denis Paperno, Anna Szabolcsi, and Jos Tellings. UCLA, pp. 427–439.

Wymark, Daniel (2018). "Relative frequencies of English NPs." Ms., UCLA.

Zimmermann, Thomas Ede (1993). "Scopeless quantifiers and operators." In: *Journal of Philosophical Logic* 22, pp. 545–561.

Zuber, Richard (2004). "A class of non-conservative determiners in Polish." In: *Lingvisticae Investigationes* 27.1, pp. 147–165.

— (2010). "Semantic constraints on anaphoric determiners." In: *Research on Language and Computation* 8.4, pp. 255–271.

— (2018). "Weak conservativity." Ms under review.

Author Index

Arnauld, Antoine, 1

Bach, Emmon, 23
Barwise, Jon, 27, 34, 36
Beghelli, Filippo, 84
Bell, John L., 144
Ben-Shalom, Dorit, 121, 126, 136
Boole, George, viii, 1–3
Boolos, George, 16, 30, 63
Burgess, John, 63
Burnett, Heather, 80

Carnielli, Walter, 63
Chomsky, Noam, vii, 1–3, 36
Cole, Peter, 132
Cooper, Robin, 27, 34

De Saussure, Ferdinand, 2
De Swart, Henriëtte, 90, 95
Dekker, Paul, 126

Ebbinghaus, Heinz-Dieter, 13
Emonds, Joseph, 117
Enderton, Herbert B., 13, 74, 147
Epstein, Richard, 63
Etchemendy, John, 109

Faltz, Leonard, 78

Fauconnier, Gilles, 36, 64, 68
Flum, Jörg, 13
Frege, Gottlob, 1

Gabbay, Dov, 36
Givant, Steven, 149
Grätzer, George, 72

Halmos, Paul, 149
Heim, Irene, 31
Hermon, Gabriella, 132
Hintikka, Jaako, 36
Hodel, Richard E., 5
Huang, C. T. James, 132

Jackendoff, Ray, viii, 3, 91
Jeffrey, Richard, 16, 63
Jelinek, Eloise, 23
Jesperson, Otto, 1
Johnsen, Lars G., 87
Jonsson, Bjarni, 150

Keenan, Edward L., 13, 23, 27,
    30–32, 40, 46, 48, 52, 55, 56, 58,
    62, 69, 78, 84, 85, 98, 103, 109,
    110, 114, 115, 117, 118, 121,
    126, 128, 130, 133, 134, 137,
    145, 149

Subject Index

a-free, 130, 131

absolute adjective, 28, 29, 81–83

AI, *see* automorphism invariant

antisymmetric, 71, 142

(argument) free, 130

atom, 58, 86, 87, 99, 104–106, 114,
    129, 133, 135, 143–147, 149, 150
    definition, 86, 144

atomic, 6, 57, 78, 86, 87, 91, 92,
    94, 95, 99, 103, 104, 106, 108,
    129, 143, 146, 150–152

atomic formula, 12–15

atomless, 144, 145, 151

automorphism, 105–115, 117

automorphism invariant, 107

autonomy of syntax, 66

bijective, 21, 47, 50, 106

boolean lattice, 47

boolean algebra, 1, 47, 71, 72, 85,
    94, 146

boolean closure, 85, 86

boolean compounds, viii, 13, 15,
    21, 27, 30, 35, 40, 69, 74–79, 85,
    124

boolean connectives, viii, 2, 7, 14,
    20, 36, 37, 76

boolean lattice, 65, 71–73, 86,

91–93, 103–106, 112, 113, 129,
    139, 143–147, 149–152
    definition, 71

boolean operations, 25, 75, 78, 79,
    85, 94, 99, 133, 148–150

bounded, 3, 62, 71, 72, 103, 114,
    128, 138, 144, 150

ca-free, 92, 94

cardinal, 21, 26–30, 33, 37, 48, 68,
    69, 100

cardinal comparative, 21, 35, 36,
    43

cardinality, 25, 84, 99, 103–105,
    108, 115, 129, 139, 149

co-cardinal, 32

co-intersective, 32

compactness, 11, 16, 38

complement, 25, 45–48, 50, 51,
    56, 57, 63, 65, 67, 72, 73, 78,
    80, 84, 92–94, 97, 108, 128, 132,
    142, 148

complete homomorphism, 78, 92,
    148–150

completeness, 11, 16, 146

conservative, 59, 83–85, 87–90, 95,
    98, 99, 104, 108, 121, 123

conservativity, 60, 83, 85, 87, 90,